Fish Diseases

NIPA® GENX ELECTRONIC RESOURCES & SOLUTIONS P. LTD.
New Delhi-110 034

About the Authors

Dr. B.K. Das, Director attained his Ph.D. in 1998 from Orissa University of Agriculture and Technology and later on completed Post-Doc at FRS Marine Lab, Aberdeen, Scotland (2006-2007). An influential figure in fisheries science, He has served as President of the Inland Fisheries Society of India and a member of the Professional Fisheries Graduates Forum (PFGF). His research contributions include research on Hilsa and river dolphin conservation, Antimicrobial Resistance (AMR) in fisheries, and cage culture diversification, improving Hilsa fish stocks across the Farakka Barrage on the Ganga River, and ranching in Hilsa upstream. He has worked on fisheries development in Bihar's Kothia Maun and major river valley projects like the Sardar Sarovar Reservoir and Narmada in Gujarat. The innovative "Waste to Wealth" project explores using black soldier fly larvae and silkworm pupae in collaboration with the Insect Institute and Central Silk Board. With more than 32 awards and honours, his notable recognitions include the Jawaharlal Nehru Award for postgraduate research (1999), the Lal Bahadur Shastri Young Scientist Award (1999-2000), the Dr. M.S. Swaminathan Award for Best Indian Fisheries Scientist (2011), the Krushi Ratna Award from Orissa Krushak Samaj (2016), and the prestigious Rafi Ahmed Kidwai Award for Outstanding Research in Agricultural Sciences (2020). In recent years, He received Lifetime Achievement Awards from VDGood Technology and the International Academy of Science and Research in 2023. He has extensive international exposure with institutions like World Fish, IUCN, FAO, and the World Bank, and countries like Malaysia, Bangladesh, Switzerland, Germany, Canada, and Australia. Dr. Das is associated with many collaborative projects with global organizations including GIZ, TWAS, NACA, and SAARC. With over 495 international publications, his scholarly output includes 37 books, 27 book chapters, 5 technical bulletins, 15 radio talks, and 29 documentary films. They have also guided 25 Ph.D. and 35 Master's students, along with postdoctoral and international students. His technological innovations include the commercialization of seven products, such as the CIFRI PEN HDPE®, CIFRI GI CAGE®, and CIFRI CAGEGROW®, registering four designs with the Indian Council of Agricultural Research (ICAR) and seven patents. He also signed 11 MoUs with government departments, three for commercial ventures, and several for consultancy, research, and academic collaborations. His leadership in fisheries science and numerous contributions to both national and international projects reflect a commitment to innovation and sustainable development in the fisheries sector.

Dr. Vikash Kumar, currently working as a senior scientist at a research institute (ICAR, Central Inland Fisheries Research Institute, Kolkata, India), completed his bachelor's in fisheries science (B.F.Sc.) at the College of Fisheries, Central Agricultural University, India. He obtained his master's in fisheries science (M.F.Sc.) degree in Fish health specialization from ICAR-Central Institute of Fisheries Education (ICAR-CIFE), India. Later, he moved to Belgium and completed his PhD at the Laboratory of Aquaculture & Artemia Reference Center (ARC), Ghent University, Belgium. During his study period, he received Government of India funding, based on his performance in class, to pursue bachelor's, master's and PhD programs. He has learned Transcriptomic, Proteomic, Cellular biology, Histology, immunohistochemistry, electron microscopy, and other techniques. He has published over 50 research papers in peer-reviewed journals, 10 book chapters, and > 20 popular articles. He has also presented the scientific outputs of his research at international congresses (World Aquaculture Conference, European Aquaculture, etc.). He received two patents (Method for producing synergistic microbial composition to promote fish health and disease protection; Means to detect whether acute hepatopancreatic necrosis disease-causing Vibrio parahaemolyticus is virulent or non-virulent). He has supervised five master's students (Mr. Bipul Kumar Dey, Bangladesh; Ms. Thi Ngoc Phuong Tran, Vietnam; Mr. Alexander Goossens 2020, Belgium; Angana Mondal 2024, India and Saikat Barui 2024, India). He served as an associate editor and member of the editorial board for several reputed journals, including Frontiers, MDPI, Aquaculture, Fish and Shellfish Immunology, Microbiology Letters, etc. He was awarded the Best Scientist Award (ICAR, Government of India), Best Speaker awards (College of Fisheries, India; Indian Virological Society; Conference Mind; Asian Fisheries Society Indian Branch and Society of Fisheries and Life Sciences), Young Scientist Award (College of Fisheries, India) and International Achievement Award (Society of Fisheries and Life Sciences).

Volume 04 : The Protocol Series

Fish Diseases Diagnosis and Treatment

Basanta Kumar Das
Director
ICAR-Central Inland Fisheries Research Institute
Barrackpore, Kolkata-700120
West Bengal, India

Vikash Kumar
Senior Scientist
ICAR-Central Inland Fisheries Research Institute
Barrackpore, Kolkata-700120
West Bengal, India

NIPA® GENX ELECTRONIC RESOURCES & SOLUTIONS P. LTD.
New Delhi-110 034

NIPA® GENX ELECTRONIC
RESOURCES & SOLUTIONS P. LTD.

101,103, Vikas Surya Plaza, CU Block
L.S.C. Market, Pitam Pura, New Delhi-110 034
Ph : +91-11-43860225, Mob.: +91 9717133558, 9540816132
E-mail: newindiapublishingagency@gmail.com
Website: www.nipaersources.com

Print ISBN: 978-93-58877-90-8
ebook ISBN: 978-93-58876-60-4

Composed and Designed by NIPA®.

Preface

The challenge of feeding a growing global population while minimizing environmental impacts is particularly pressing in developing countries. The demand for animal protein is expected to rise, exacerbating environmental challenges like droughts and floods. Increasing fish and seafood consumption, particularly through aquaculture can serve as a potential solution. The aquaculture industry, crucial for livelihoods and food security, faces risks from disease outbreaks due to intensified production. Environmental factors like water quality and climate change further heighten these risks, making fish more susceptible to bacterial pathogens, which can cause significant economic and ecological damage.

Advances in molecular diagnostic techniques for detecting fish pathogens offer promising solutions for managing these risks. These techniques, which include DNA sequencing and PCR can allow for faster and more accurate diagnosis of diseases, in comparison to traditional culture methods. However, widespread adoption of these methods requires validation and integration into routine diagnostic practices.

Disease prevention in aquaculture focuses on maintaining a healthy environment through proper sanitation, chemo-prophylaxis, vaccination, and environmental manipulation. Treatment methods vary depending on the situation, including the use of chemicals in water, feed, or direct application to fish. The success of fish production depends on the health of species, making disease monitoring and management crucial.

Despite progress in diagnostic techniques, there are a few of challenges, such as variability of disease symptoms, involvement of multiple pathogens, and human error in diagnosis. To ensure the effectiveness of new molecular methods, large-scale validation and trials are needed. These efforts will be essential for advancing fish health management and sustaining aquaculture production in the face of growing global demand and environmental pressures.

The initiative by ICAR-CIFRI to publish the book entitled "Fish Diseases: Diagnosis and Treatment" with an innovative approach is commend-able. We believe this book will set a significant benchmark in addressing fish diseases and their prevention.

Authors

Contents

1

Background

A key challenge for the years to come is feeding a rapidly growing human population while lowering the impact of food production on the environment. This is particularly true for low and middle-income countries where the demand for animal protein is likely to rise and further increase several folds in existing environmental changes (e.g., droughts, floods, extensive wildfires) have in recent years led to major food crises. To achieve these goals, food production not only needs to be increased, but most of all, good husbandry practices must be followed to reduce its negative impacts on the environment. Several studies have suggested that shifting the human diet towards increased consumption of fish and seafood could be a solution to the need for protein that would sustain human and environmental health. In fact, fish and seafood consumption is forecast to increase by 27% on the horizon of 2030, mostly sustained by the aquaculture sector, which is expected to grow by 62% during the same period. The aquaculture industry contributes significantly to livelihood, food security, and poverty alleviation, with over 100 million people estimated to rely on aquaculture for their living. Fish is a cheap source of animal protein, and the current per capita consumption of fish in India is around 9 kg per annum, which is compared to the 11 kg recommended by the World Health Organisation (WHO). Presently, the contribution of the fisheries sector to the gross domestic product (GDP) and agriculture GDP has been estimated to be 1.2 and 4.2%, respectively.

However, due to the increase in global demand, the pressure for intensification of production underpins environmental health and creates stressful conditions that make an ideal ground for disease outbreaks. For instance, dissolved oxygen in water and pH levels play a critical role in the overall health and growth of fish, as well as the success of aquaculture operations. It can affect various biological and chemical processes, such as nutrient availability, gas exchange, osmoregulation, and the toxicity of certain substances, which can result in stress, illness, and even death if not properly managed (Figure 1). Currently, infectious disease is the single most devastating problem in shrimp and fish culture and presents an ongoing threat to other aquaculture sectors. In Asian countries, losses due to disease problems have been estimated to be more than

US$ 3000 million. Most disease problems may be caused by bacteria, viruses, fungi, or parasites. Annual losses due to diseases in carp culture in Andhra Pradesh are estimated at Rs 40 million. Viral disease in shrimp culture has become a stumbling block, and the economic loss was estimated at Rs 9450 million in 1995-96. So suitable and approximate health management measures involving the development of diagnostics and vaccines are necessary for disease diagnosis, forecasting, and planning of sound aquaculture practices.

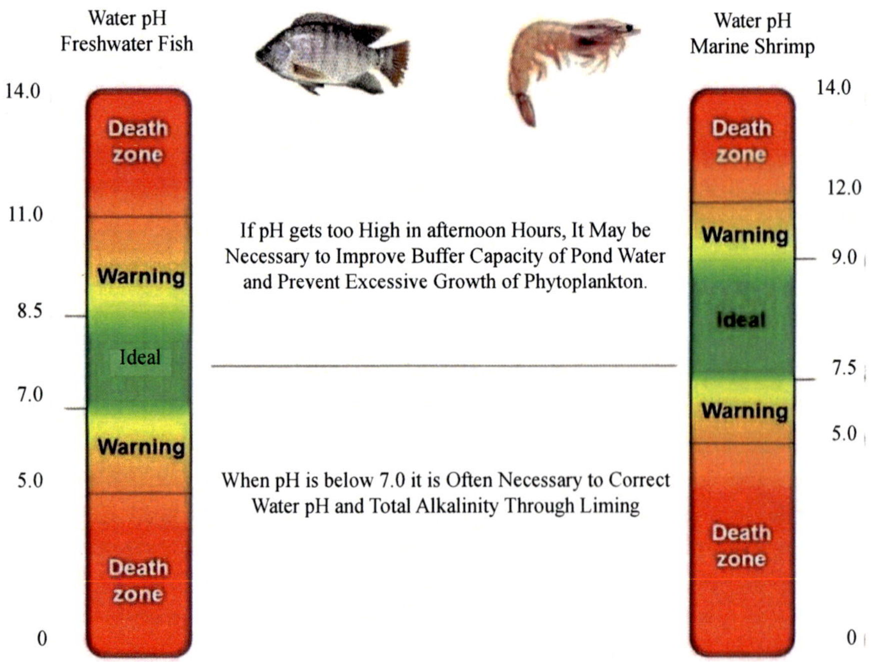

Figure 1. Illustrates the crucial role of water pH in maintaining the health of fish and shrimp in aquaculture. Understanding and managing water pH levels is essential for ensuring the well-being of aquatic species and the success of aquaculture operations.

Health and Aquatic Animals

Unlike other farm and harvesting situations, where the animals and plants are visible, aquatic animals require more attention in order to monitor their health. They are not readily visible except under tank-holding conditions, and they live in a complex and dynamic environment. Likewise, feed consumption and mortalities may be equally well hidden underwater. Unlike the livestock sector, aquaculture has a wide range of diversity in species cultured, farming environment, nature of containment, intensity of practice and culture system used. The range of diseases found in aquaculture is also varied, some with low or unknown host specificities and many with non-specific symptoms.

Disease is now recognized as one of the most important challenges facing the aquaculture sector.

The complexity of the aquatic ecosystem obscures the distinction between health, sub-optimal performance, and disease. Diseases in aquaculture are not caused by a single event but are the end result of a series of linked events involving the interactions between the host (including physiological, reproductive, and developmental stage conditions), the environment, and the presence of a pathogen (Figure 2). Under aquaculture conditions, three factors are particularly important in affecting the host's susceptibility: stocking density, innate susceptibility, and immunity (natural/acquired). Environment includes not only the water and its components (such as oxygen, pH, temperature, toxins, and wastes) but also the kind of management practices (e.g., handling, drug treatments, transport procedures, etc.). Pathogens may include viruses, bacteria, parasites, and fungi; diseases may be caused by a single species or a mixture of different pathogens. The introduction of infectious diseases is another major concern in aquaculture. As in livestock, the aquaculture and fisheries sector will continue to face increasing global exposure to disease agents as it intensifies trade in live aquatic organisms and their products.

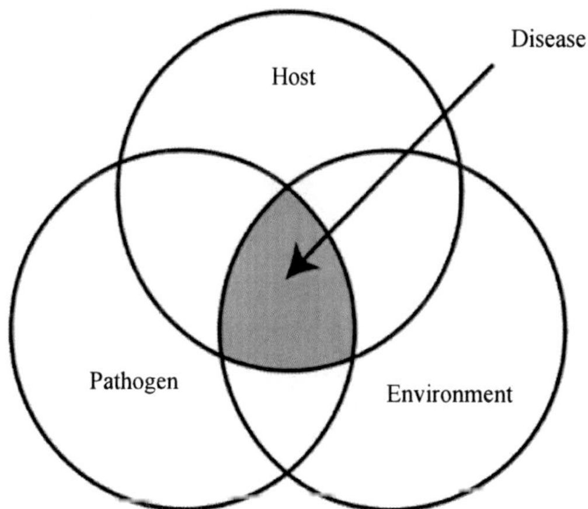

Figure 2. Disease is the outcome of pathogens, environment, and host fish.

In the context of aquaculture, disease may be broadly defined as any condition that leads to sub-optimal production. A definition from the medical literature states that a disease is the sum of the abnormal phenomena displayed by a group of living organisms in association with a specified common characteristic or set of characteristics by which they differ from the norm of their species in such a way as to place them at a biological disadvantage. Aquatic animal diseases can be caused by:

1. Primary pathogens without the involvement of environmental stress;
2. Primary pathogens as a result of environmental stress;
3. Opportunistic pathogens as a result of environmental stress; and
4. Poor environmental conditions without the involvement of pathogens.

The degree to which an animal is susceptible to pathogens and the occurrence of clinical diseases largely depends on three main tactics: The status of the host organism, the environment, and pathogens. When one or more factors are unfavorable, the host must adopt its physiology or behaviors to compensate. These adaptive responses, stress, impair normal physiological functioning and reduce the host's chance of survival (Figure 3). In particular, chronic stress lowers the resistance of fish to infectious agents. Specific examples of things that can cause stress (stressors) are listed in Table 1.

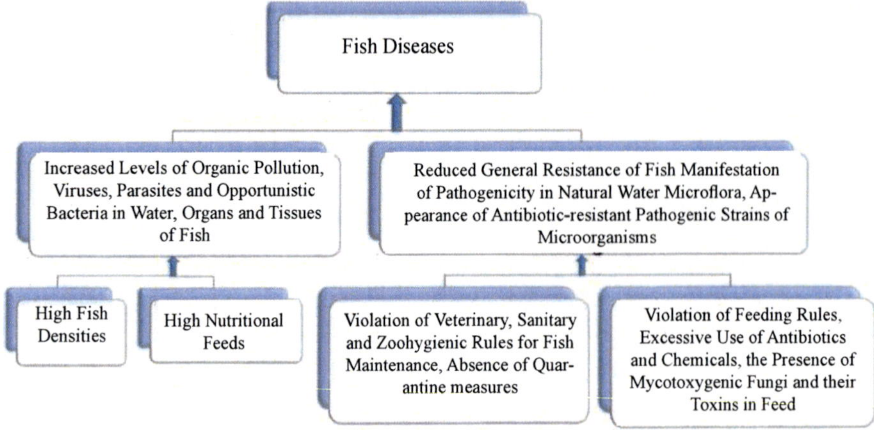

Figure 3. The factors responsible for disease development in aquaculture.

Table 1. Different types of stressors in the aquaculture system

Chemical stressors	**Biological stressors**	**Physical stressors**	**Procedural stressors**
i. Poor water quality-low dissolved oxygen ii. Pollution iii. Diet composition-types of protein, amino acids iv. Nitrogenous and other metabolic wastes accumulation of ammonia or nitrite	i. Population density crowding ii. other species of fish-aggression, territoriality, lateral swimming space requirements iii. Microorganisms-pathogenic and non-pathogenic iv. Macro organisms-internal and external parasites	i. Temperature ii. Light iii. Humidity iv. Dissolved gases v. Dissolved solids	i. Handling ii. Shipping iii. Disease treatments

What to do if Your Fish are Sick

The first and most important defenses against preventable disease losses under such complex situations are monitoring as regularly as possible and appropriate action at the first sign(s) of suspicious behavior, lesions, or mortalities. If you suspect that fish are getting sick, the first thing to do is check the water quality. Low oxygen is a frequent cause of fish mortality in ponds, especially in the summer. High levels of ammonia are also commonly associated with disease outbreaks when fish are crowded in vats or tanks. Separate extension fact sheets are available that explain oxygen cycles, ammonia cycles, and management of these water quality problems. In general, check dissolved oxygen, ammonia, nitrite, and pH during a minimum water quality screen associated with a fish disease outbreak. The parameters of significance include total alkalinity, total hardness, nitrate (saltwater systems), and chlorine (if using city water).

Ideally, daily records should be available for immediate reference when a fish disease outbreak occurs. These should include the dates fish were stocked, size of fish at stocking, source of fish, feeding rate, growth rate, daily mortality, and water quality. This information is needed by the aquaculture specialist working with you to solve your fish disease problem. Good records, a description of behavioural and physical signs exhibited by sick fish, and results of water quality tests provide a complete case history for the diagnostician working on your case. If you decide to submit fish to a diagnostic laboratory, you should collect live, sick fish, place them in a freezer bag (without water), and ship them on ice to the nearest facility. Small fish can be shipped alive by placing them in plastic bags that are partially filled (30 percent) with water. Oxygen gas can be injected into the bag prior to sealing it. An insulated container is recommended for shipping life, bagged fish as temperature fluctuations during transit are minimized. In addition to fish samples, a water sample collected in a clean jar should also be submitted.

Diseases caused by bacterial pathogens pose significant risks to the aquaculture sector and global food security worldwide and result in the loss of primary productivity and biodiversity that negatively impact the environmental and socio-economic conditions of affected regions. Climate change further increases disease outbreak risks by altering bacterial pathogen evolution, changing host-pathogen interactions and vector physiology, and facilitating the emergence of new strains of bacterial pathogens, which in turn can break down host resistance. For instance, pathogens can shift and increase the spread of bacterial diseases in aquaculture conditions, while fish's immune response becomes weaker in changing anthropogenic climate conditions, making the host more susceptible to pathogens. Hence, the urgency of advancing fundamental knowledge of emerging bacterial pathogens and host biology, ecology and evolution is a key step forward to unraveling the complexities of disease incidence and severity and the socio-economic impacts on disease burden that could have devastating consequences in aquaculture production system, impacting food production and security, ecosystem sustainability and social conflicts.

Role of Diagnostics in Aquatic Animal Health Management and Disease Control

Diagnostics play two significant roles in aquatic animal health management and disease control. As described above, some diagnostic techniques are used to screen healthy animals to ensure that they are not carrying infection at sub-clinical levels by specific pathogens. This is most commonly conducted on stocks or populations of aquatic animals destined for live transfer from one area or country to another. Such screening provides protection on two fronts: (a) it reduces the risk that animals are carrying few, if any, opportunistic agents that might proliferate during shipping, handling, or change of environment, and (b) it reduces the risk of resistance or tolerant animals transferring a significant pathogen to a population which may be susceptible to infection. The second role of diagnostics is to determine the cause of unfavorable health or other abnormality (such as spawning failure, growth, or behavior) in order to recommend mitigating measures applicable to the particular condition. This is the most immediate and clearly recognized role of diagnostics in aquatic animal health.

Accurate diagnosis of a disease is often incorrectly described as complicated and costly. This may be the case for some of the more difficult-to-diagnose diseases or newly emerging diseases. Disease diagnosis is not solely a laboratory test. A laboratory test may confirm the presence of a specific disease agent, or it may exclude its presence with a certain level of certainty. Incorrect diagnosis can lead to ineffective or inappropriate control measures (which may be even more costly). For example, a "new" disease agent may get introduced to a major aquaculture-producing area, or the animals may all die in shipment/during handling. Disease diagnostics should be made as a continuum of observations starting on the farm and, in fact, commencing prior to the disease event. The different levels of disease diagnostics that can be undertaken when investigating a disease situation are discussed in the section below.

Levels of Diagnostics

Disease Diagnostic Guide is built on a framework of "three levels" of diagnostics. Table 2 below outlines the diagnostic activities at each level, who is responsible, and the equipment and training required. It should be noted that none of the levels function in isolation but build on each other, each contributing valuable data and information for optimum diagnoses. Level 1 provides the foundation and is the basis of Levels II and III since findings using higher level(s) can only be meaningfully interpreted in conjunction with observations and results obtained from lower levels.

Level I (farm/production site observations, record-keeping, and health management) is strongly emphasized throughout the Asia Diagnostic Guide as this forms the basis for triggering the other diagnostic levels (II and III). Level II includes the specializations of parasitology, histopathology, bacteriology, and mycology, which require moderate capital and training investment and which, generally speaking, cannot be conducted at the farm or culture site. Level III comprises the types of advanced diagnostic specialization which require significant capital and training investment. As the reader will note, immunology and biomolecular techniques are included in Level III, although field kits are now being developed for farm or pond-side use (Level I) as well as use in microbiology or histology laboratories (Level II). These efforts are a good indication that technology transfer is now enhancing diagnostics, and with solid quality control and field validation, it is certain that more Level III technology will become field accessible in the near future.

Table 2. Diagnostic Levels, Associated Requirements and Responsibilities

Level	Activity	Work requirements	Responsibility	Technical requirements to support activities
I	Observation of animal and environment Gross clinical examination	Knowledge of normal (feeding, behaviour, growth) of stock. Frequent / regular observation of stock. Regular, consistent record-keeping and assistance (Levels II, III). Maintenance of records – including fundamental environmental information. Knowledge contacts for health diagnosis. Ability to submit and/or preserve representative specimens for optimal diagnosis (Levels II, III).	Farm worker/manager. Fishery extension officers. On-site veterinary support. Local fishery biologists.	Field keys. Farm record keeping formats. Equipment lists Model clinical observation sheets. Pond/Site record sheets. Preservation/transportation guidelines for Levels II/III diagnoses. Model job descriptions/skill requirements. Asia Diagnostic Guide for Aquatic Animal Diseases
II	Parasitology Bacteriology Mycology Histopathology	Laboratories with basic equipment and Personnel trained/experienced in aquatic animal pathology. Keep and maintain accurate diagnostic and laboratory case records. Ability to preserve and storage specimens for optimal Level III diagnoses. Knowledge of/ contact with different areas of specialisation within Level II. Knowledge of who to contact for Level III diagnostic assistance.	Fish biologists/ technicians. Aquatic Veterinarians. Parasitologists/ technicians. Mycologists/ technicians. Bacteriologists/ technicians. Histopathologists/ technicians.	Model laboratory record-keeping system. Protocols for preservation/ transport of samples to Level III. Model laboratory requirements/ equipment/ consumables lists. Model job descriptions/ skill lists. Access to Level II and Level III specialist expertise. Asia Diagnostic Guide for Aquatic Animal Diseases OIE Diagnostic Manual for Aquatic Animal Diseases Regional General Diagnostics Manuals.

III	Virology Electron microscopy Molecular biology Immunology	Highly equipped laboratory with highly specialised and trained personnel. Keep and maintain accurate diagnostic and laboratory case records. Preserve and store specimens. Maintenance of contact with people responsible for sample submission.	Virologist/ technician. Ultrastructural histopathologist/ technicians. Molecular biology scientists/ technicians.	Model laboratory requirements/ equipment/ consumables lists. Model job descriptions/ skill requirements. Contact information for reference laboratories. Protocols for preservation of samples for consultation/ validation. *Asia Diagnostic Guide for Aquatic Animal Diseases* *OIE Diagnostic Manual for Aquatic Animal Diseases* *General molecular and microbiology diagnostic references*

A diverse range of Gram-positive and Gram-negative bacteria have been associated with diseases of marine fish worldwide. Overall, greater attention has been focused on aquaculture rather than wild stocks. The pathogens have been associated with a wide range of clinical manifestations, including ulcerations, swellings, erosions, and hemorrhagic septicaemias. The diagnostic procedures have been summarised in Figure 4. Unlike other veterinary and medical counterparts, diagnosis centers on the identification of the pathogen rather than the immediate control of the disease. Inevitably, this process takes longer than, for example, a trip to the human physician. Here, diagnosis may result initially from consideration of the gross clinical signs of the patient, leading to the rapid implementation of a treatment regime. Certainly, fish disease diagnosis has undergone a transformation. Since the end of the twentieth century, there has been a move away from the traditional histology and culture-dependent approaches. The latter involved the acquisition of cultures and their time-consuming identification. Subsequently, attention has been focused on culture-independent techniques, notably those embracing developments in molecular biology. A typical example is the sequencing of the 16S rRNA gene, which does not need intact, viable bacterial cells. The advantage of these culture-independent approaches is speed and accuracy; the bacteria may be studied and identified regardless of whether or not they may be grown in the laboratory. Moreover, there is a very high level of specificity, and this is important when diagnosing disease.

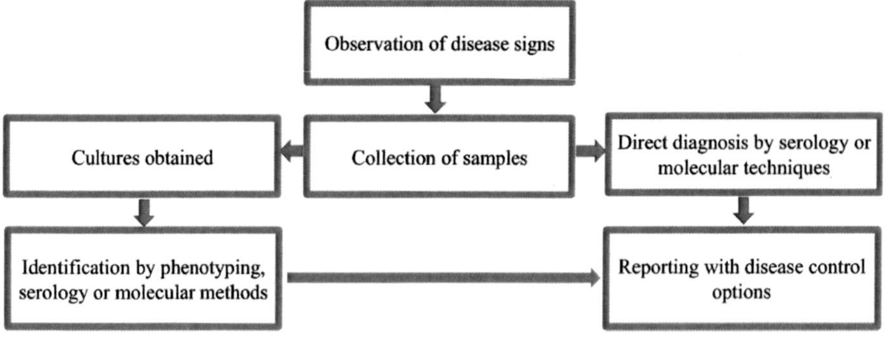

Figure 4. Steps in disease diagnosis of fish

The development of molecular tools to detect fish pathogens is well-advanced. The study of emerging diseases will also benefit from these advances. Many alternative methods exist for some diseases, each targeting a slightly different region of a gene or using different methods such as probes, sequencing, or PCR. While this progress in the application of DNA technology is very welcome, it is now time for those dealing with a molecular diagnosis of fish disease to prove the utility of these methods and to bring them into the portfolio of methodologies acceptable and available to fish health workers.

Considering the adoption of diagnostic techniques, scientific advisors to regulatory authorities should critically review all available peer-reviewed methods and examine the degree to which each has been validated and applied on a routine basis, as will be required in diagnostic laboratories. The time has now come for validation and wide-scale application of diagnostic methods for pathogen identification in fish tissue as well as inter-laboratory ring tests. Following this, a period of use alongside existing techniques is likely to establish the congruence of the different techniques (Figure 5).

One of the most important aspects of maximizing the effectiveness of the three diagnostic levels is ensuring that Level I diagnosticians have access to and know how to contact Level II and III support (and at what cost) and vice versa. Level III diagnostic support is usually based on referrals, so it has little contact with field-growing conditions. They, therefore, need feedback to ensure any diagnosis (and actions recommended) are relevant to the aquatic animal production situation being investigated.

Thus, the baseline aim for initiating diagnostic capability is Level I. Confirmatory diagnostics, or second opinion, where required, can be obtained by referral until such capabilities are developed locally. The period required to develop Level II and/or Level III diagnostic infrastructures usually depends on the disease situations being faced and tackled by Level I diagnosticians in the area/country and the resources available. Where there are few problems, there is little incentive to build diagnostic capability. This is a vulnerable position and strong links with Level II and/or III diagnostics are good precautionary measures and strongly promoted under the regional program – especially for introductions of live aquatic animals into a relatively disease-free area.

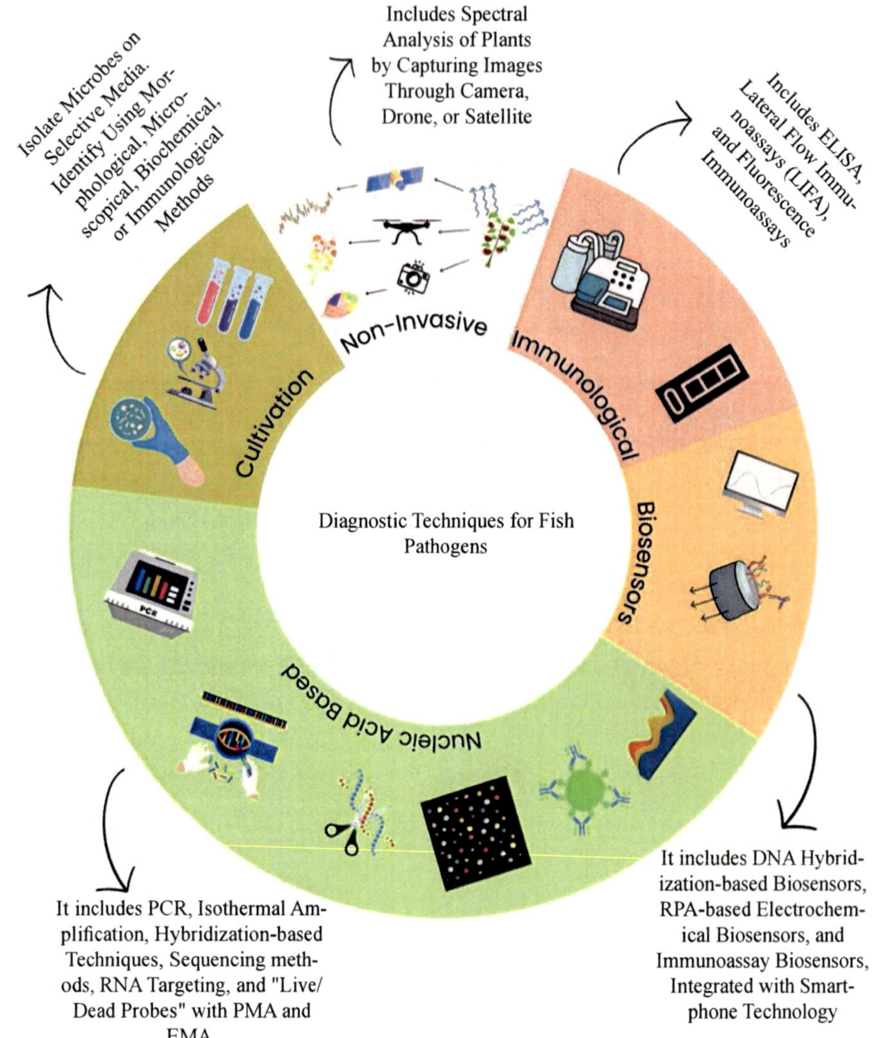

Figure 5. Fish disease diagnostic techniques

Principles of Fish Disease Control and Prophylaxis Treatment

Even though many developments have occurred in the area of fish health management, there is still belief in the principle that "prevention is better than cure." Maintaining the health of the environment, one can. To protect the occurrence of disease to a greater extent. Different prophylaxis measures and chemical applications are used to control fish disease.

Disease prevention can be achieved by (1) proper sanitation of the environment establishments and appliances, (2) chemo-prophylaxis, (3) vaccination, and (4) manipulation of the environment. The culture system should be properly prepared to create a pathogen-free and hygienic congenial environment in which the species can be stocked. Clearance of aquatic weeds, de-siltation, quick lime application, etc., may help to maintain the ecosystem's productivity. Careful selection of quality stocking material that has better resistance to disease and a faster growth rate helps reduce losses and achieve higher production. Before planting fish on new grounds, they should undergo quarantine checks so as to eliminate any undesirable traits or disease agents. Diseased fish that carry virulent pathogens and are unmanageable should be destroyed to prevent the spread of the disease.

Different Methods of Treatment

Treatment may be applied in very many ways, and the particular type of treatment to be applied is to be decided based on the specific situations encountered. There are three ways of applying the treatment, e.g., (1) adding chemicals to the water, (2) adding chemicals to the feed, and (3) administering chemicals directly to individual fish. Different types of treatment are:

Dip: Fish are placed in a hand net and dipped into a concentrated solution of the drug for one to three minutes or less. **Flush**: Here, a concentrated solution of the drug or chemical is added to the inlet and allowed to pass through the tank or the raceway with the water flow. **Short bath**: Here, the required amount of chemical or drug is added directly to the rearing or holding unit and left for a specified period of time. **Indefinite bath**: This is most widely used in ponds where a low concentration of a chemical or drug is applied and is allowed to dissipate naturally. **Feeding**: Treatment through feed aims at reaching the chemical into the stomach of the sick fish at the proper dosage. Injection: Large and valuable fish may be treated by injecting the medicine into the body. **Topical application**: Sometimes, valuable fish may be treated by direct topical application of the drug.

2

Bacterial Diseases

A. Disease Name: Bacterial gill disease or Gill Rot

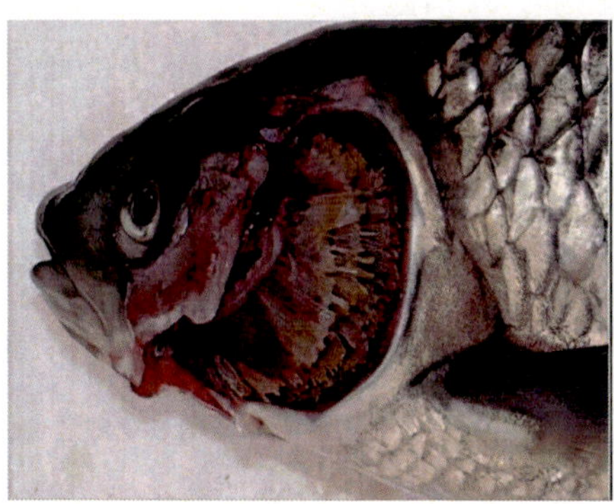

Clinical Symptoms

- Gasping for air
- Discoloured gills with trapped materials
- Swollen gills with enhanced mucous secretion
- White to grey spots on the gills
- Diseased fish gather near the outlet of the pond
- Fish become lethargic and remain on the surface waters

Causative Agent: *Flavobacterium branchiophilum, Cytophaga* spp., *Flexibacter* spp.

Host (Common sp.)

- Goldfish
- Koi carp
- Cichlids
- Fighter fish
- Gourami fish

Season/ Growth Stage of Animal for Outbreak: Spring and early summer months.

Treatment Measures

- Chloramine-T for 1 hour at 8 ppm
- Potassium permanganate at 100mg/L for 30 seconds
- Hyamine-1622 (98.8%) at 1-2 ppm for 1 hour
- Purina Four Power at 3-4 ppm (1-h flush treatments)
- Diquat (8.4-16.8 ppm or 2-4 ppm of formulation)
- CuSO4.5H2O at 1:2000 ppm (1-min bath)

B. Disease Name: Columnaris (Saddleback) Disease

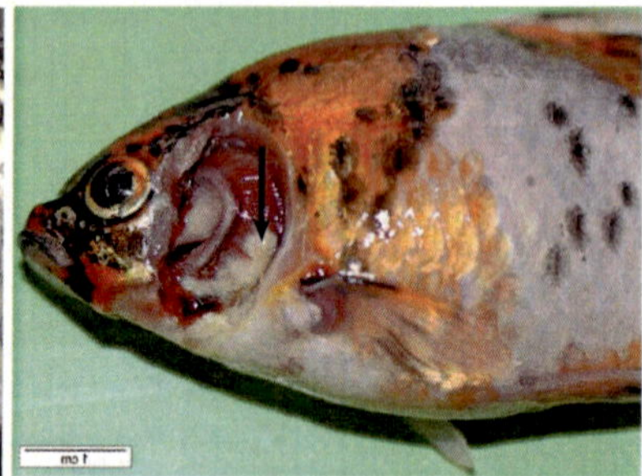

Clinical Symptoms

- White or grayish spots or patches on the head, fins, or gills
- Lesions on the back that may extend down the sides
- Moldy or cottony-looking lesions on the mouth
- Frayed fins
- Fish may become lethargic, exhibit a loss of appetite, and hang at the surface

Causative Agent: *Flavobacterium columnare*

Host (Common sp.)

- Most freshwater fishes (cultured and wild) are considered susceptible

Season/ Growth Stage of Animal for Outbreak: Winter

Treatment Measures

- Quarantine the fish in a separate hospital tank
- Treatment with copper sulfate ($CuSO_4$) bath at 37 mg/L for 20 minutes
- Addition of copper sulfate to pond water at 0.5 mg/L
- Potassium permanganate ($KMnO_4$) at 2 mg/L

- Treatment with oxytetracycline (10 gm/100 kg)
- FURAN-2 at 1 packet per 10 gallons

C. Disease Name: Edwardsiellosis

Clinical Symptoms

- Exophthalmia
- Ascites
- Hernia
- Hemorrhagic septicaemia
- Ulcerative abscesses in internal organs
- Abnormal swimming
- Fibrinous peritonitis

Causative Agent: *Edwardsiella tarda*

Host (Common sp.)

- Both freshwater and marine fish, especially in the warm environment

Season/ Growth Stage Of Animal For Outbreak: Summer (25°C-30°C) to post-monsoon season.

Treatment Measures

- Antibiotic treatment (based on sensitivity test)
- Improvement of water quality
- Stress reduction (housing, hiding places, population density and group composition in the tank e.g.)
- Treatment with probiotics and ascorbic acid

D. Disease Name: Tail Rot & Fin Rot

Clinical Symptoms

- Erosions, discoloration and disintegration of fins and tails
- Fins and tail appear to melt away or discolour
- Blood on the edges of fins
- Ulcers with grey or red margins
- Lethargic and cloudy eyes
- Erratic swimming
- Loss of appetite
- Fish may settle at the bottom of the tank.

Causative Agent: *Aeromonas hydrophila, Pseudomonas* spp., *Cytophaga* spp., *Haemophilus* spp.

Host (Common sp.)

- Cichlids
- Fighter fish
- Livebearer fish
- Gourami fish
- Goldfish
- Koi Carp

Season/ Growth Stage of Animal for Outbreak: Summer (25°C-30°C) to post-monsoon season.

Treatment Measures

- Tetracycline at 3-4 gm/100 L for 2-3 days followed by water change
- Mild dip treatment with KMnO4 (0.002%)
- Mild dip treatment with copper sulphate (0.05 %)
- Chloromycetin at 40 mg in 5 L of water

- Antibiotics (oxytetracycline, **erythromycin, streptomycin, nalidixic acid etc.)**

E. Disease Name: Aeromoniasis

Clinical Symptoms

- Enlarged eyes (exophthalmos)
- Accumulation of fluids in the abdomen (ascites)
- Renal dropsy (kidney damage)
- Rupture of minor blood vessels
- Distended abdomen in acute cases
- Tissue necrosis in many organs
- Hemorrhagic septicaemia
- Ragged fins

Causative Agent: *Aeromonas hydrophila, Aeromonas veronii, Aeromonas sobria, Aeromonas caviae*

Host (Common sp.)

- Freshwater fish such as catfish and bass, and many species of tropical or ornamental fish.

Season/ Growth Stage Of Animal For Outbreak: Spring to early summer season.

Treatment Measures

- Potassium permanganate (KMnO4), at a rate of 2 to 4 parts per million
- Vitamins C and E at 1000 mg/kg of feed
- Levamisole at 5 mg/kg feed on five occasions at three days intervals
- Terramycin medicated feed at a rate of 25 to 37.5 milligrams of active drug per pound of fish per day for a period of 10 days

F. Disease Name: Dropsy

Clinical Symptoms

- Distended abdomen
- Scale protrusion
- Inflammation of intestine
- Haemorrhagic ulcer on skin and fins
- Bulging eyes
- Behavioral changes including lethargy
- Increased respiratory rate

Causative Agent: *Aeromonas hydrophila, Pseudomonas fluorescens*

Host (Common sp.)

- Dropsy can occur in all freshwater fish, including koi, goldfish, guppies and bettas

Season/ Growth Stage of Animal for Outbreak: Throughout the year.

Treatment Measures

- Potassium permanganate dip at 5 ppm dip for 2 minutes
- Oxytetracycline at 50 mg/day/kg of fish orally with feed for 10 days
- 1 teaspoon of salt for every gallon of water

G. Disease Name: Bacterial Kidney Disease

(*Source:* Colorado Parks & Wildlife/John Drennan)

Clinical Symptoms

- Darkened skin
- Pale gills
- Swelling of kidney, heart, spleen and liver
- Unilateral or bilateral exophthalmia
- Ocular lesions
- Blebs and blisters on the epidermis with white or yellowish hemorrhagic fluid
- Abdominal distention due to the accumulation of ascitic fluid in the abdominal and pericardial cavities

Causative Agent: *Renibacterium salmoninarum*

Host (Common sp.)

- All salmonids are considered susceptible and the disease usually occurs in fish 6 months or older, i.e., juvenile and adult fish

Season/ Growth Stage Of Animal For Outbreak: Outbreaks can occur throughout the year but are more prevalent during summer

Treatment Measures

- Prolonged treatment with clindamycin, erythromycin, penicillin G, spiramycin and lincomycin

H. Disease Name: Furunculosis

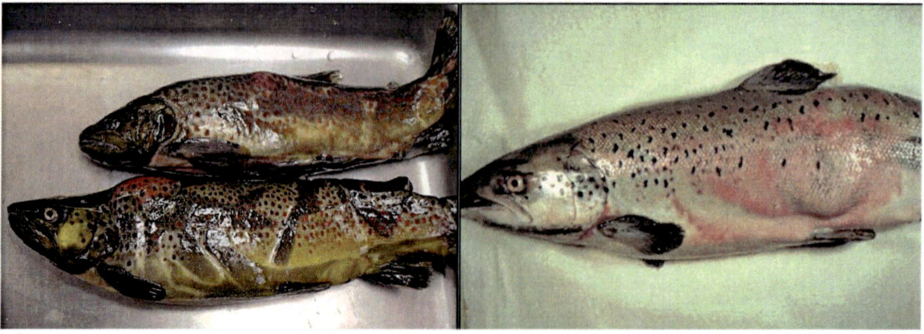

(*Source:* Robert Durborow, 2012)

Clinical Symptoms

- Blisters on epithelial and muscular tissues
- Bleeding from the skin, bases of fins and anus
- Gradual destruction of pectoral fins and oral cavity
- Paling out of gills and alteration in body colour
- Lethargic swimming or swimming just below the surface
- Enlarged spleen and focal necrosis of the liver
- Stomach filled with mucus, blood and sloughed epithelial cells
- Congested intestine

Causative Agent: *Aeromonas salmonicida*

Host (Common sp.)

- All salmonids and eels are believed to be susceptible

Season/ Growth Stage of Animal for Outbreak: Mostly occurs during the summer season.

Treatment Measures

- Terramycin (oxytetracycline) should be added to feed at the rate of 3.0 g/l00 lb fish, administered daily for 10 d to affected fish
- Sulfamerazine should be administered at the rate of 5-10 g/100 lb fish and fed for 10 or 15 consecutive days.
- Dip treatment for one minute in 1:2000 copper sulphate solution for 3-4 days

I. Disease Name: Cold Water Disease (CWD)

(*Source:* Craig Banner)

Clinical Symptoms

- Loss of appetite
- Eroded fin tips
- Gradual destruction of pectoral fins and oral cavity
- Erosive skin and muscle lesions
- Focal necrosis in the spleen, liver, and kidneys
- Exophthalmia
- Abdominal distension with increased volumes of ascites

Causative Agent: *Flavobacterium psychrophilum*

Host (Common sp.)

- Affects a broad host-species range of fishes that inhabit cold, fresh waters

Season/ Growth Stage of Animal for Outbreak: Occurs predominately at water temperatures of 16 °C and below, and is most prevalent and severe at 10 °C and below.

Treatment Measures

- **A 60-minute bath treatment in furanace at 10-15 mg/L.**
- Aquaflor at 10 mg/ kg of fish for 10 days
- Terramycin at 3.75 g/ 45.4 kg of fish for 10 days
- Oxytetracycline at 10-50 mg/L or quaternary ammonium compounds at 2 mg/L
- Sulfamethazine administered at concentrations of 220 - 440 mg/kg in starter diets and at concentrations of 110 mg/kg in pelleted feeds

J. Disease Name: Exophthalmia (Popeye)

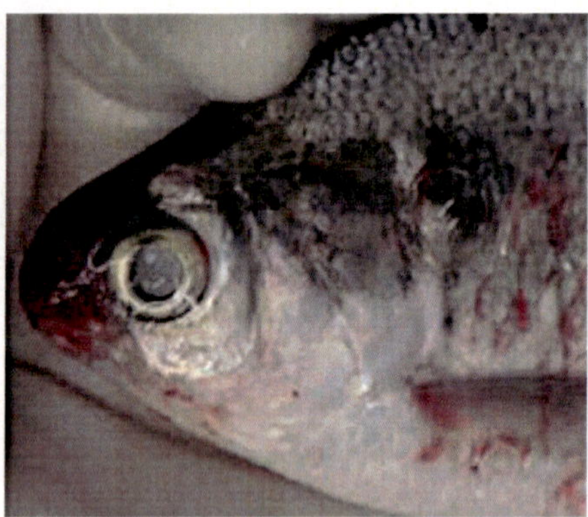

Clinical Symptoms

- Protrusion of one or both eyeballs
- Stretching of the eye socket
- Discoloration or blood in the eyeball
- Rupture of the eyeball
- Cloudiness of the eyeball
- Hiding or other behavioral changes
- Loss of appetite
- Swollen body and clamped fins

Causative Agent: *Aeromonas punctate*

Host (Common sp.)

- Affects a broad host-species range of fishes

Season/ Growth Stage of Animal for Outbreak: Throughout the year

Treatment Measures

- Chloromycetin (8-10mg/l) bath for one hour for 2-3 days
- Treatment with Epsom salt

K. Disease Name: Vibriosis

(*Source:* Mohamad *et al.*, 2019)

Clinical Symptoms

- Lethargy, loss of appetite, skin and fin ulcerations with body discolouration
- Characterised by external skin lesions, haemorrhages, and septicaemia

Causative Agent: *V. salmonicida, V. anguillarum, V. ordalli, V. harveyi, V. alginolyticus, V. vulnificus, V. parahaemolyticus, V. mimicus and Photobacterium damselae subsp. damselae* cause significant impacts on cultured fish and shellfish species.

Host (Common sp.)

- Affects a broad host-species range of fish, shrimp and molluscan species

Season/ Growth Stage of Animal for Outbreak

- Late summer
- Fish or shellfish of any age group are susceptible to the infection but, young animals are more prone

Treatment Measures

- Antibiotics (20-50 mg/kg) such as oxytetracycline, tetracycline, quinolones, nitrofurans, potentiated sulfonamides, trimethoprim, sarafloxacin, flumequine, and oxolinic acid have been used to treat vibriosis.

L. Disease Name: Mycobacteriosis

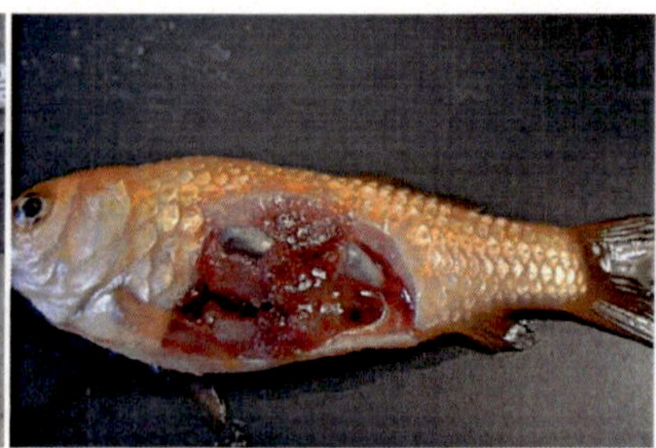

(*Source:* Gauthier and Rhodes, 2009)

Clinical Symptoms

- Scale loss, dermal ulceration, emaciation, spinal defects and ascites
- Uncoordinated swimming, abdominal swelling, loss of weight, skin ulceration, white nodule formation as granuloma in liver, kidney, spleen

Causative Agent: *Mycobacterium chelonae, M. marinum,* and *M. fortuitum.*

Host (Common sp.)

- Affects a broad host-species range of fishes

Season/ Growth Stage Of Animal For Outbreak: Winter season

Treatment Measures

- Antibiotics (20-50 mg/kg) such as tigecycline, tobramycin, clarithromycin and amikacin have been used to treat Mycobacteriosis.

M. Disease Name: Streptococcosis

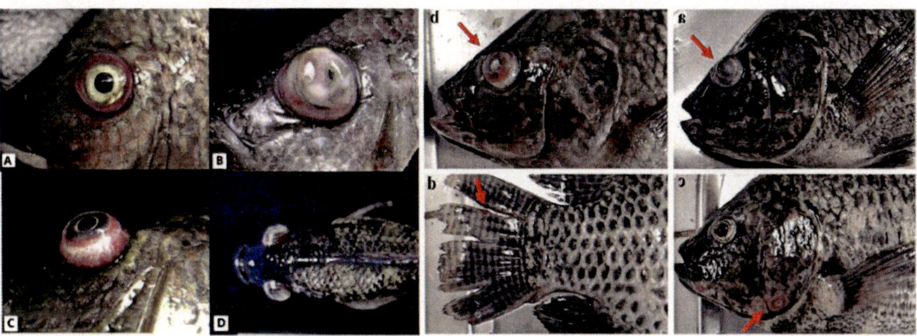

(*Source:* Wang *et al.*, 2022)

Clinical Symptoms

- Cloudy eyes
- Exophthalmos
- Focal, ulcerative skin lesions along the caudal peduncle area

Causative Agent: *Streptococcus iniae*

Host (Common sp.)

- *Seriolae quinqueradiata*, *Epinephelus lanceolatus*, *Lutjanus* spp.

Season/ Growth Stage of Animal for Outbreak: Late winter and early spring, and lower in summer and autumn

Treatment Measures

- Antibiotics (20-50 mg/kg) such as florfenicol, erythromycin, doxycycline and oxytetracycline have been used to treat Streptococcosis.
- Disinfecting agents can dissolve in water (copper sulfate and formalin) to avoid infection.

3

Viral Diseases

A. Disease Name: Spring viremia of carp

Clinical Symptoms

- External haemorrhages
- Pale gills
- Ascites
- Destruction of tissues in the kidney, spleen, and liver
- Excessive mucus secretion
- Skin hemorrhages with edema of the tissues
- Skin ulceration around the mouth and base of the fins

Causative Agent: *Rhabdovirus*

Host (Common sp.)

- Common carp
- Koi carp
- Crucian carp
- Silver carp
- Bighead carp
- Grass carp
- Goldfish

Season/ Growth Stage of Animal for Outbreak: Spring season

Treatment Measures

- Iodophore treatment of eggs
- Physical and chemical disinfection of ponds and farm equipments
- Inactivation with 3% formalin for 5 minutes, chlorine (500 ppm), iodine (0.01%), NaOH (2% for 10 minutes)
- UV irradiation (254 nm) and gamma irradiation (103 krads)

B. Disease Name: Viral Haemorrhagic Septicemia

Clinical Symptoms

- Bulged eyes
- Distended abdomen due to oedema in the peritoneal cavity
- Bruised-looking reddish tints to the eyes, skin, gills and fins
- Hemorrhages of the internal organs, skin and muscle
- Darkening of the skin
- Lethargy

Causative Agent: *Viral haemorrhagic septicaemia virus*

HOST (Common sp.)

- Rainbow trout, in which VHSV can cause severe disease outbreaks, are traditional hosts of the virus, but many other freshwater and marine fish species are also susceptible to the disease.
- VHS affects over 40 different species of fish. This includes several important recreational, sport and commercial fish species such as salmon, trout, yellow perch, sunfish, muskellunge, walleye, northern pike and several minnow species

Season/ Growth Stage of Animal for Outbreak: Typical outbreaks occur in spring and autumn at temperatures generally below 14°C resulting in an acute to chronic disease.

Treatment Measures

- No therapies are currently available

C. Disease Name: Lymphocystis

(*Source:* fitzfishponds.com)

Clinical Symptoms

- Irregular, moderate sized, nodular, wart- like growths on the fins, skin or gills.
- Pop-eye

Causative Agent: *Lymphocystis disease virus*

Host (Common sp.)

- Teleosts
- Killifishes
- Gouramies
- Sunfishes
- Clownfish
- Drum
- Sea bass

Season/ Growth Stage of Animal for Outbreak: Mostly prevalent during winter

Treatment Measures

- There is no good treatment that will speed up recovery from this disease
- Culling (removing) infected fish from a population may help reduce overall loads of viruses in the system as well as the infection rate

D. Disease Name: Tilapia Lake Virus Disease

Clinical Symptoms

- Darkening or lightening of body colour
- Skin erosion resulting in haemorrhagic dermal lesions
- Scale protrusion
- Exophthalmos (popeye)
- The opacity of the eye lens (cataract)
- Abdominal distension due to fluid or enlargement of spleen and other organs

Causative Agent: *Tilapia lake virus*

Host (Common sp.)

- *Sarotherodon galilaeus*
- *Oreochromis niloticus*

Season/ Growth Stage Of Animal For Outbreak: Late summer to monsoon

Treatment Measures

Dip-treatment in

- NaOCl at 10 ppm for 10 min
- H_2O_2 (300 ppm for 10 min)
- Iodine (2.5 ppm for 10 min)
- Formaline (80 ppm for 60 min)
- Virkon (5000 ppm for 1 min)
- Chlorine (20 ppm for 30 min)
- PVD (50 ppm for 30 min)

E. Disease Name: Tilapia Parvovirus Disease

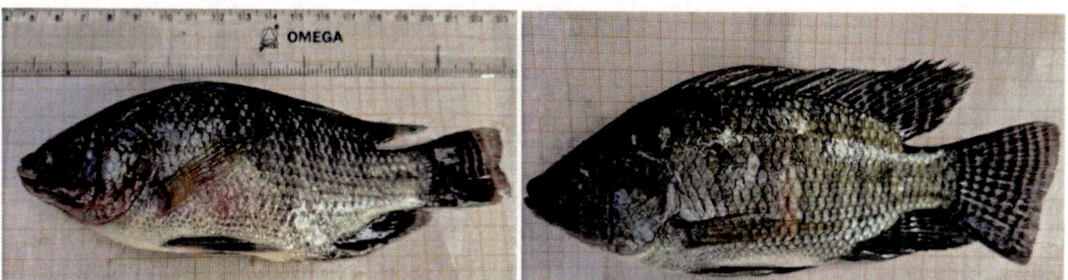

Clinical Symptoms

- Lethargy
- Change in swimming behaviour
- Anorexia
- Hemorrhages on the body surface, lower jaw, anterior abdomen, and fin bases
- Exophthalmia
- Ocular lesions

Causative Agent: *Tilapia parvovirus*

Host (Common sp.)

- Tilapia

Season/ Growth Stage Of Animal For Outbreak: Prevalent during summer season (28°C and 30°C)

Treatment Measures

- No therapies are currently available

F. Disease Name: Carp pox virus

(*Source:* Rahmati-Holasoo *et al.*, 2020)

Clinical Symptoms

- Mucoid to smooth waxy multifocal-to-coalescing, slightly-raised-to-nodular, milky-white, cutaneous growths were observed
- Epidermal papillomas in common carp and other cyprinids
- Necrotic cells with pyknotic nuclei and hypereosinophilic cytoplasm were frequent

Causative Agent: *Cyprinid herpesvirus 1* (CyHV-1)

Host (Common sp.)

- Common carp (*Cyprinus carpio),* Goldfish (*Carassius auratus*) and grass carp (*Ctenopharyngodon idella*)

Season/ Growth Stage Of Animal For Outbreak: Prevalent during spring season (22°–25.5°C)

Treatment Measures

- No therapies are currently available

G. Disease Name: Herpesviral hematopoietic necrosis

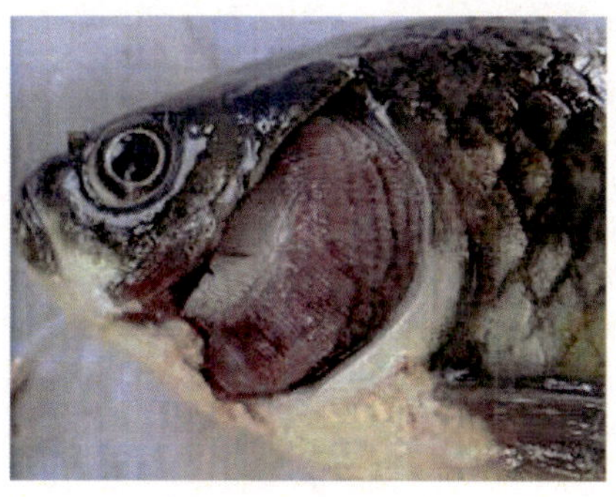

(*Source:* Sahoo *et al.*, 2016)

Clinical Symptoms

- Apathy, anorexia, pale gills and pale skin patches
- Enlarged kidney and distended spleen with white focal nodules
- Necrosis of hematopoietic tissues of spleen and kidney

Causative Agent: *Cyprinid herpesvirus-2* (CyHV-2)

Host (Common sp.)

- Common carp (*Cyprinus carpio)*, Goldfish (*Carassius auratus*) and grass carp (*Ctenopharyngodon idella*)

Season/ Growth Stage of Animal for Outbreak: Prevalent during spring season (22°–25.5°C)

Treatment Measures

- No therapies are currently available

H. Disease Name: Koi herpesvirus

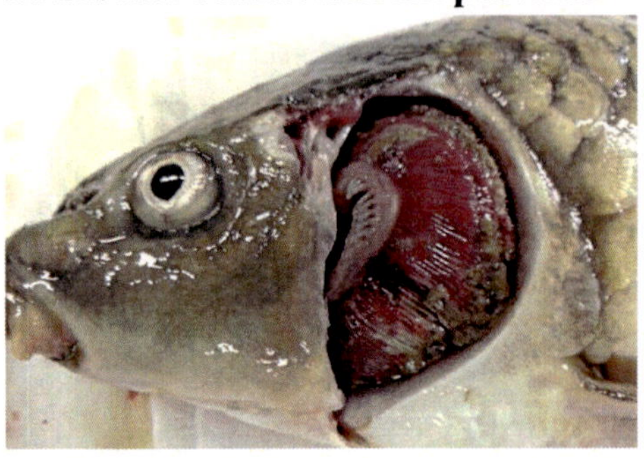

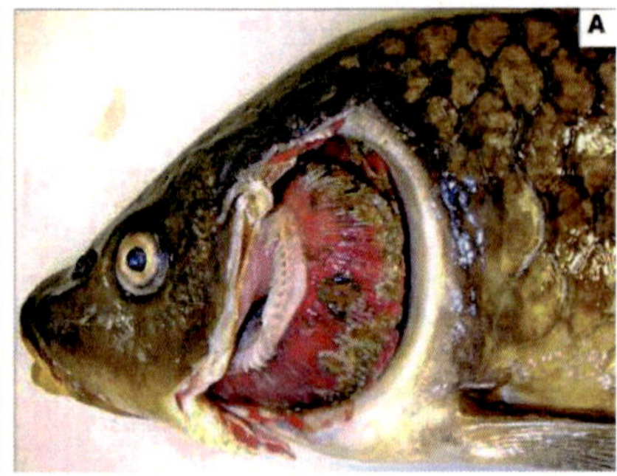

Clinical Symptoms

- Gills: Swollen, necrotic, or discolored gills, with mucus hypersecretion
- Skin: Reddened skin, fins, or tail, with lesions or ulcerations
- Eyes: Sunken eyes
- Behavior: Disorientation, erratic swimming, hyperactivity, or "hanging" with a head down

Causative Agent: *Cyprinid herpesvirus-3* (CyHV-3)

Host (Common sp.)

- Common carp (*Cyprinus carpio)*, Goldfish (*Carassius auratus*) and grass carp (*Ctenopharyngodon idella*)

Season/ Growth Stage of Animal for Outbreak: Prevalent during spring season (22°–25.5°C)

Treatment Measures

- No therapies are currently available

I. Disease Name: Viral Encephalopathy and Retinopathy (VER) or Viral Nervous Necrosis (VNN)

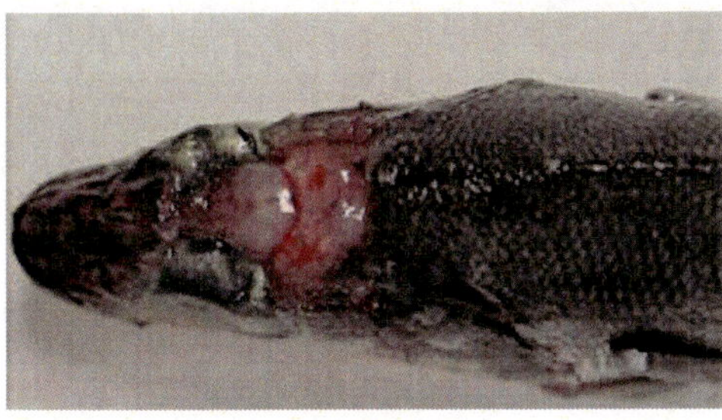

(*Source:* Vendramin *et al.*, 2013)

Clinical Symptoms

- Higher mortalities in larvae and juvenile fish (9–28 days old) although older fish
- Fish resting belly-up (loss of equilibrium)
- Sporadic protrusion of the head from the water
- Blindness, abrasions, emaciation, colour change
- Overinflated swim bladder (the only significant internal gross pathological sign)

Causative Agent: *Betanodavirus*

Host (Common sp.)

- VER has been found in at least 40 species of marine fish from 16 families

Season/ Growth Stage of Animal for Outbreak: Throughout the year

Treatment Measures

- No therapies are currently available

J. Disease Name: Infectious hematopoietic necrosis (IHN)

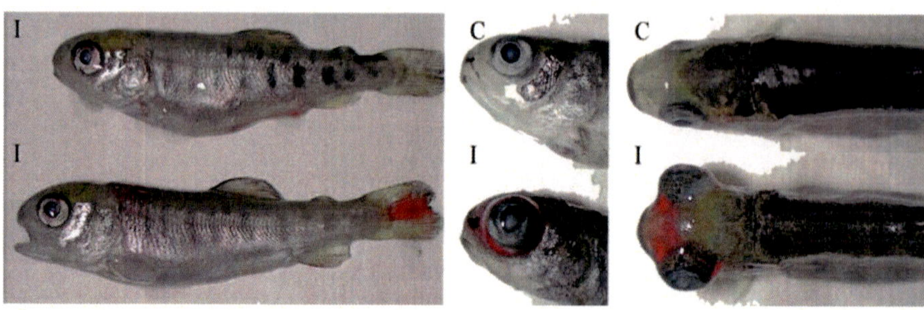

(*Source:* Romero *et al.*, 2008)

Clinical Symptoms

- Abdominal distension, exophthalmia, darkened skin and pale gills.
- Long, semi-transparent fecal casts often trail from the anus.
- Affected fish are typically lethargic, with bouts of hyperexcitability and frenzied, abnormal activity.
- Petechial hemorrhages commonly occur at the base of the pectoral fins, the mouth, the skin posterior to the skull above the lateral line, the muscles near the anus, and the yolk sac in sac fry.
- In sac fry, the yolk sac often swells with fluid. In fry less than two months old, there may be few clinical signs despite a high mortality rate.
- Surviving fish often have scoliosis.

Causative Agent: *Infectious hematopoietic necrosis virus*

Host (Common sp.)

- Trout and salmon
- Rainbow/ steelhead trout (*Oncorhynchus mykiss*), cutthroat trout (*Salmo clarki*), brown trout (*Salmo trutta*), Atlantic salmon (*Salmo salar*), and Pacific salmon including chinook (*O. tshawytscha*), sockeye/ kokanee (*O. nerka*), chum (*O. keta*), masou/ yamame (*O. masou*), amago (*O. rhodurus*), and coho (*O. kisutch*).

Season/ Growth Stage of Animal for Outbreak

- Clinical disease usually occurs when the water temperature is between 8°C (46°F) and 15°C (59°F)

Treatment Measures

- No therapies are currently available

K. Disease Name: Infectious salmon anemia (ISA)

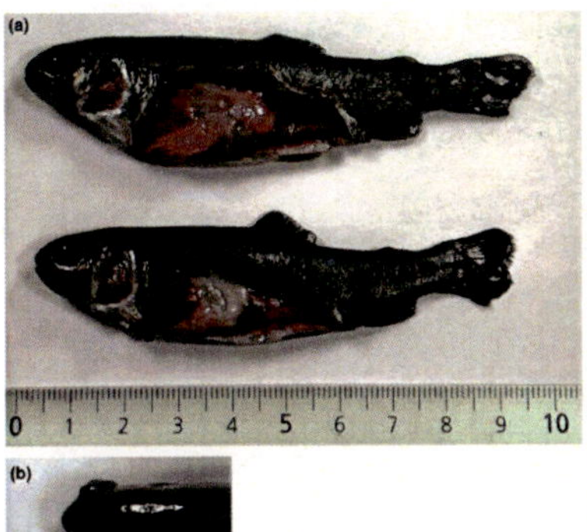

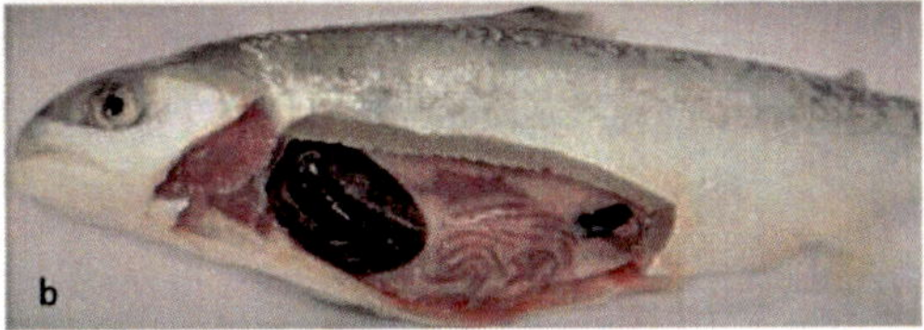

(*Source:* Biacchesi *et al.*, 2007)

Clinical Symptoms

- Pale gills (except in the case of blood stasis in the gills), exophthalmia, distended abdomen, blood in the anterior eye chamber, and sometimes skin haemorrhages especially of the abdomen, as well as scale pocket oedema
- Nutritional status is usually quite normal, but diseased fish has no feed in the digestive tract

Causative Agent: *Infectious salmon anaemia virus*

Host (Common sp.)

- Atlantic salmon (*Salmo salar*), rainbow trout (*Oncorhynchus mykiss*), Coho salmon (*Oncorhynchus kisutch*), brown trout and sea trout (*S. trutta*), pollock (*Pollachius virens*) and cod (*Gadus morhua*)

Season/ Growth Stage of Animal for Outbreak: Throughout the year

Treatment Measures

- No therapies are currently available

4

Parasitic Diseases

A. Disease Name: Larneasis

Clinical Symptoms

- Anchor worms can be seen with the naked eye
- Frequent rubbing or "flashing"
- Localised redness
- Inflammation on the body of the fish
- Tiny white-green or red worms in wounds
- Breathing difficulties

Causative Agent: *Lernaea* spp.

Host (Common sp.)

- A wide array of freshwater fishes, especially wild-caught and pond-raised species

Season/ Growth Stage of Animal for Outbreak: Most prevalent in the spring months

Treatment Measures

- Diflubenzuron, a pesticide interferes with the growth of the parasite at a dose of 0.066 mg diflubenzuron/litre
- Prolonged immersion in trichlorfon, an organophosphate is effective for ornamental fishes
- 30-minute bath with potassium permanganate at 25 mg/L

B. Disease Name: Argulosis

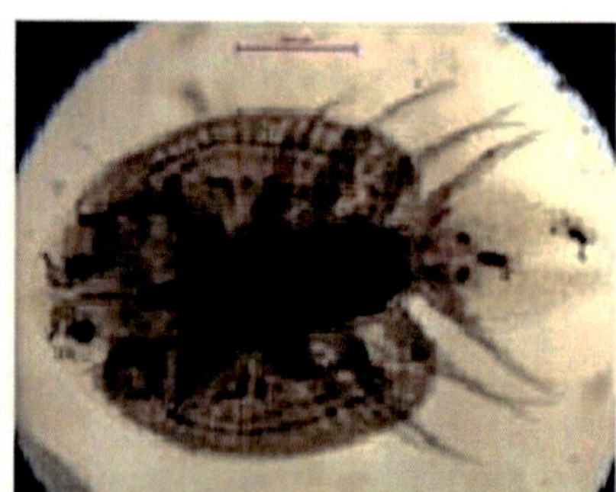

Clinical Symptoms

- Loss of scales
- Skin fissures
- Hyperplasia at the wound margin
- The wounded place by argulosis becomes red colored
- The growth of fish gets reduced and the body becomes feeble
- Increased mucus production
- Anemia

Causative Agent: *Argulus* spp.

Host (Common sp.)

- Infections are most common in wild and pond-raised freshwater fish, particularly goldfish, koi, and other cyprinids (carps and minnows); centrarchids (sunfishes) and salmonids (salmon and trout)

Season/ Growth Stage of Animal for Outbreak: The disease is more prevalent during summer seasons.

Treatment Measures

- Treating with gammexane at a concentration of 0.2 ppm repeated at weekly intervals
- Trichlorfon, dosed at 0.25–0.50 mg/L active ingredient, once a week for 4 treatments
- Potassium permanganate (10 mg/L for 30 minutes, or 1.3 mg/L applied twice over 3 days)

C. Disease Name: Icthyophthiriasis (white spot disease)

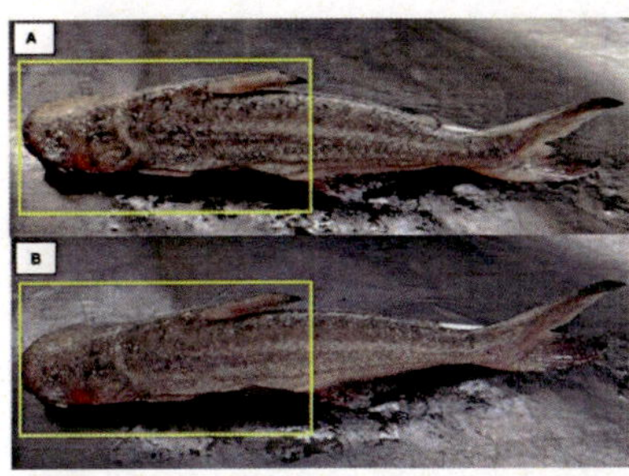

Clinical Symptoms

- The presence of small white spots on the skins or fins
- Anorexia
- Rapid breathing
- Hiding abnormally
- Resting on the bottom
- Flashing

Causative Agent: *Icthyophthirius multifilis*

Host (Common sp.)

- Infections occur across a wide array of fish species

Season/ Growth Stage of Animal for Outbreak: Throughout the year

Treatment Measures

- Copper at a dosage of 0.15-0.3 mg/L
- Aquarium salt at 3-5 gms/L for 2 weeks
- Formalin at concentrations between 100 and 250 milliliters per liter (100-250 ppm) as a short-term bath

D. Disease Name: Marine white spot disease

(*Source:* Saeed *et al.*, 2023)

Clinical Symptoms

- Small white spots, nodules, or patches on their fins, skin, or gills
- Ragged fins, cloudy eyes, pale gills, increased mucus production, or changes in skin color, and they may appear thin.
- Fish may flash (scratch), swim abnormally, hang at the surface or on the bottom, act lethargic, or breathe more rapidly as if in distress

Causative Agent: *Cryptocaryon irritans*

Host (Common sp.)

- Infections occur across a wide array of fish species

Season/ Growth Stage Of Animal For Outbreak: Temperatures between 15 and 30°C

Treatment Measures

- Ultraviolet (UV) sterilization: 280,000 μWsec/cm^2 to 800,000 μWsec/cm^2
- Ozonation
- Chemical treatment: copper sulfate pentahydrate (0.15–0.20 mg/L free copper (Cu^{2+})), Hyposalinity (Prolonged exposure to 15–16 g/L salinity or less), Chloroquine (10 mg/L chloroquine diphosphate as a prolonged bath), Formalin (16–18 g/L and 25 mg/L formalin every other day, for 4 weeks.

E. Disease Name: Whirling Disease

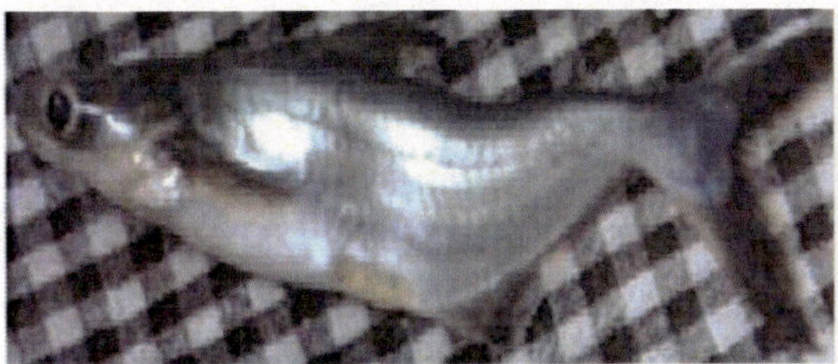

Clinical Symptoms

- Darkening of the skin from the vent to the tail (blacktail)
- Spinal curvature
- Skull deformation and shortened gill plates
- Convulsive movements
- Swimming in a whirling motion (tail chasing)
- Increased rate of breathing
- Jerking backwards movements
- Erratic, nervous darting movements until exhausted

Causative Agent: *Myxobolus cerebralis*

Host (Common sp.)

- Atlantic salmon
- Brook and Brown trout
- Chinook salmon
- Cutthroat trout
- Golden trout
- Rainbow trout
- Sockeye salmon

Treatment Measures

- There is no known cure or vaccine for whirling disease

F. Disease Name: Black-spot Disease

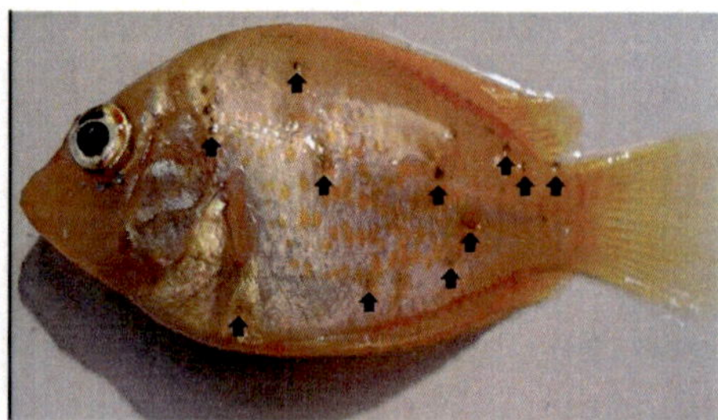

(*Source:* López-Hernández *et al.*, 2023)

Clinical Symptoms

- Infested fish exhibit black, raised nodules in the skin which are often less than 1mm in diameter
- Rubbing or scratching against substrate or tank decor
- Lethargy
- Faded color
- Loss of appetite

Causative Agent: *Uvulifer ambloplitis, Crassiphiala bulboglossa, Apophallus donicus, Cryptocotyle lingua*

Host (Common sp.)

- Salmonids and other freshwater and marine fish worldwide

Treatment Measures

- No method of control is available for the elimination of this problem
- Removal of the resident molluscan population and aquatic birds around it
- Use of molluscides

G. Disease Name: Trichodiniasis

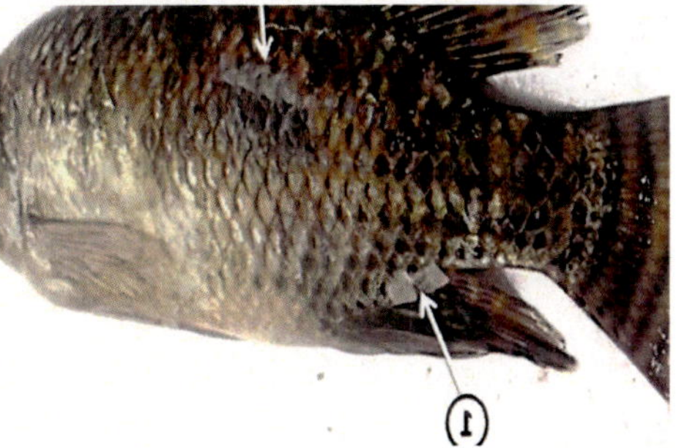

(*Source:* El-Zahraa and Ahmed, 2015)

Clinical Symptoms

- Affected fish produce excess mucus and develop white cast to skin
- Focal areas of reddening
- Lays near substrate to rub against the surface
- Lethargic
- Swelling of gills
- Sluggish and shows asphyxia

Causative Agent: *Trichodina* spp.

Host (Common sp.)

- Affects the production of several fish species throughout the world and has deleterious effects on cichlid (e.g., tilapia) production in particular

Treatment Measures

- Application at a concentration of 2-3% solution of sodium chloride (salt) bath
- Formalin at 30 ppm in pond water, in morning hours
- Mixture solution of $CuSO_4$ and $FeSO_4$ (5:2) dispersed evenly over the whole infected pond

H. Disease Name: Gyrodactylosis/ Dactylogyrosis

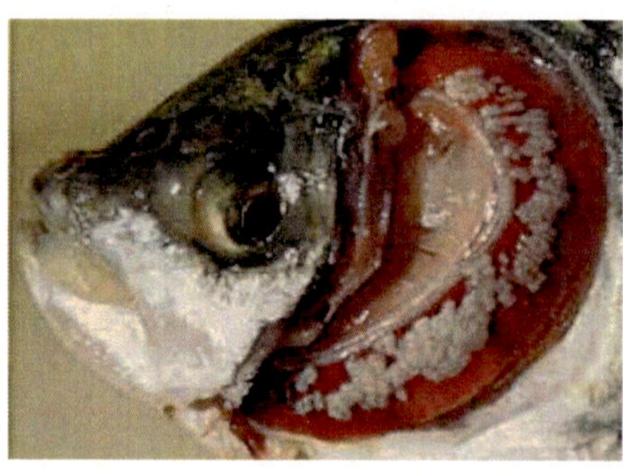
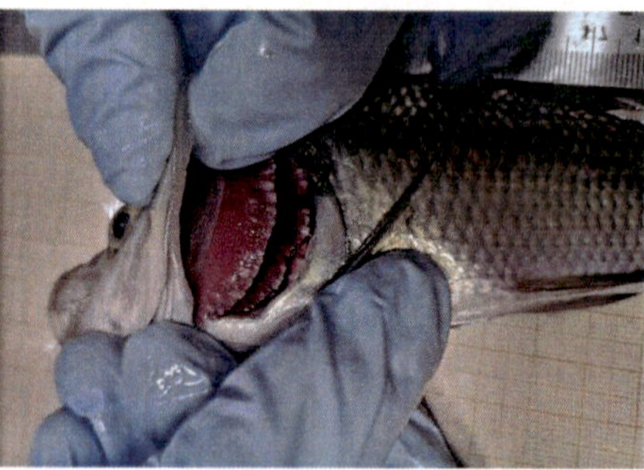

Clinical Symptoms

- Faded gills
- Dropping of scales
- Restlessness
- Mottled and necrotic skin

Causative Agent: *Gyrodactylus* spp., *Dactylogyrus* spp.

Host (Common sp.)

- The genus *Gyrodactylus* has many species in Eurasia and North America that parasitize both marine and freshwater fish
- The genus *Dactylogyrus* is found worldwide parasitizing mostly cyprinids in freshwater

Treatment Measures

- Bath for 3-5 minutes in 200-250 ppm formalin
- Application of KMnO4 in pond water at 5 ppm
- Bath in 10 ppm KMnO4 for 1-2 hours
- Trichlorfon (0.25-3.0 mg/1 liter water for 3 days)
- Formalin: 250-330mg/1 liter for 1-30 miutes by bath method

I. Disease Name: Ichthyobodiasis (Costiasis)

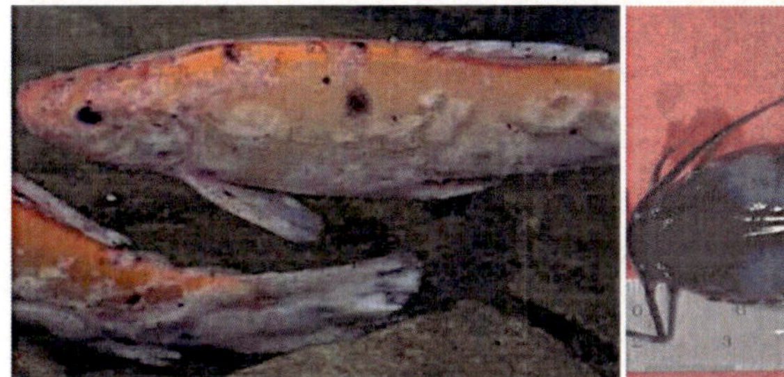
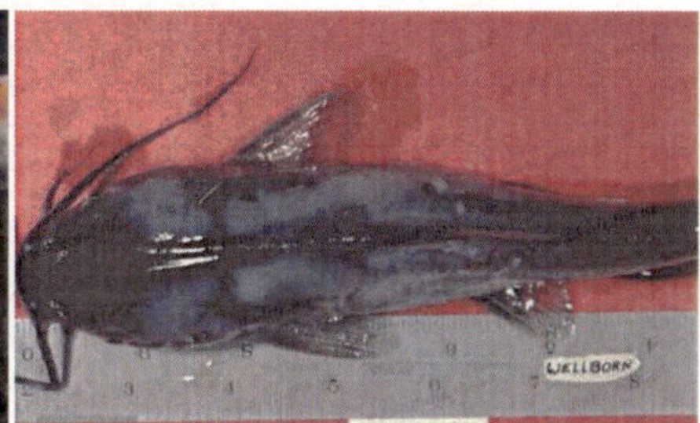

(*Source:* krishimala.com)

Clinical Symptoms

- This parasitic infection affects the skin and gills of the fish.
- The skin of the infected fish looks steel-grey in color and produces a blue or grey colored mucus.
- The fish may swim near the water surface to gulp air and may also rub against objects.

Causative Agent: *Ichthyobodo necator*

Host (Common sp.)

- Costia has been detected in many fish species, including farmed and wild-caught fish.

Treatment Measures

- Formalin, salt, potassium permanganate, or copper sulphate treatment is suggested
- A formalin bath (0.025%, 1 h) for infected salmon was used to control ectoparasites in hatcheries

J. Disease Name: Amyloodiniosis (velvet disease)

(*Source:* reefsaltwateraquarium.com)

Clinical Symptoms

- Visible, small white spots appear on the fins and skin.
- Infected fish gasp rapidly at the surface, or congregate on the bottom of the container where they swim erratically and lose equilibrium.

Causative Agent: *Amyloodinium ocellatum*

Host (Common sp.)

- Parasite infect several fish species, including farmed and wild-caught fish.

Treatment Measures

- Formalin: 36% formaldehyde at a concentration of 4mg/L for 7 hours or 51mg/L for 1 hour
- Hydrogen peroxide: 75 and 150mg/L for 30min and repeated after 6 days
- Chloroquine phosphate: 5–10 mg/L water

5

Fungal Diseases

A. Disease Name: Saprolegniasis (Cotton wool fungus)

Clinical Symptoms

- White/brown cotton-like foci on the surface of the skin and/or gills
- Peritonitis
- Extensive haemorrhage
- Necrosis
- Adhesions

Causative Agent: *Saprolegnia* spp.

Host (Common sp.)

- All freshwater fish species, incubating eggs and other lower aquatic vertebrates/invertebrates worldwide are susceptible to saprolegniasis

Season/ Growth Stage of Animal for Outbreak: Winter to early spring

Treatment Measures

- Maintaining sufficient oxygen concentrations (4 to 5 ppm)
- Treating with formalin (1-2 mL/L for 15-minute bath treatment or 0.23 mL/L for 60-minute bath treatment)
- Malachite green (1-2 mg/L bath treatment for 30-60 minutes or 0.1 mg/L prolonged immersion)

B. Disease Name: Ichthyosproridosis (Swimging disease)

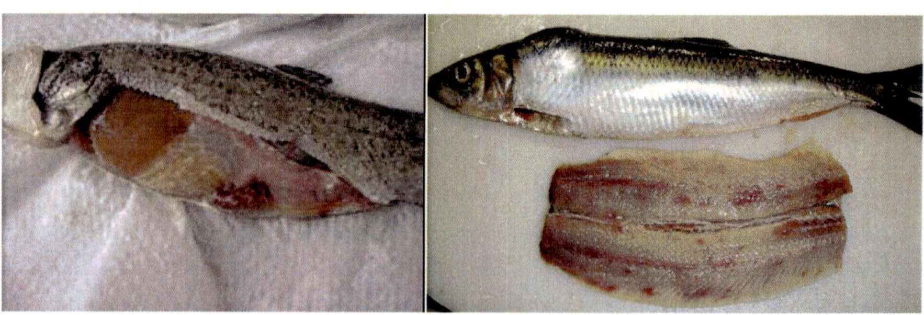

(*Source:* Hershberger, 2012)

Clinical Symptoms

- Rough or granulomatous skin
- Nodules found in heart, liver, kidney, and brain and filled with cellular debris and fungus
- Infection of the swim bladder

Causative Agent: *Ichthyophonus hoferi*

Host (Common sp.)

- Ichthyophonus has infected a wide range of freshwater, marine and anadromous fish (80 species), amphibians and reptiles worldwide

Season/ Growth Stage of Animal for Outbreak: Throughout the year

Treatment Measures

- Salt treatment

C. Disease Name: Aphanomycosis (Epizootic Ulcerative Syndrome (EUS))

Clinical Symptoms

- Small haemorrhageic spots over the body ultimately turn into big ulcers of the size of a coin with sloughing of scales and degeneration of epidermal tissue
- Individuals of fish species get infected at a time within a very short period

Causative Agent: *Aphanomyces invadans, Aphanomyces piscicida*

Host (Common sp.)

- Around 76 species of fish have been confirmed to be infected by the disease worldwide

Season/ Growth Stage Of Animal For Outbreak: More prevalent in winter

Treatment Measures

- Maintenance of water pH at around 8.0 steadily throughout the winter season
- Disinfection of pond water with $KMnO_4$ at 1.0ppm
- Dip treatment at 500 ppm $KMnO_4$ to the highly infected fishes
- Dilution of 130ml "CIFAX" and spread over one bigha water surface
- Dip treatment of infected fish at 3-4% common salt and release immediately

D. Disease Name: Branchiomycosis (Gill Rot Disease)

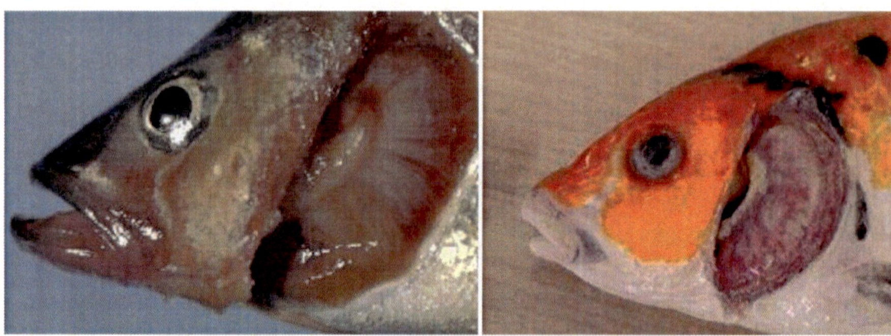

Clinical Symptoms

- Infarctions of the gills, anorexia, and marbling appearance on gills due to disturbance of circulation.
- The gill appears necrotic, eroded and pale.
- Infected fish exhibit respiratory symptoms and a loss of equilibrium.

Causative Agent: *Branchiomyces sanguinis, Branchiomyces demigrans*

Host (Common sp.)

- Branchiomycosis has been reported in a broad taxonomic range of fish species

Environmental parameters/ Season/ Growth Stage of Animal for Outbreak

- More prevalent in spring (20°C to 25°C)
- Low pH (5.8 to 6.5), low dissolved oxygen, or a high algal bloom

Treatment Measures

- Strict sanitation and disinfection are essential for disease control.
- Dead fish should be collected and daily and burned or deeply buried.
- Ponds with enzootic branchiomycosis should be dried and treated with calcium oxide (quicklime) or 2 to 3 kg copper sulphate per hectare.
- Diseased fish can be treated with malachite green at 0.1mg/l for extended periods or 0.3mg/l for 12 hours.
- Transportation of infected fish areas to non-infected areas must be prevented.
- An increase in water supply helps in the control of disease.
- Stress factors must be avoided.
- Regulating the feeding rate during warm weather.

E. Disease Name: Achlya Infection

(*Source:* Panchai *et al.*, 2014)

Clinical Symptoms

- Loss of appetite, the darkness of skin, fin rot, eroded scales and bloody wounds on the body and in the caudal region
- Yellow to grey cottony mass on the skin of flank and caudal peduncle
- Listlessness, erratic swimming and rising near water surfaces or resting on the bottom

Causative Agent: *Achlya oblongata*

Host (Common sp.)

- All life stages of freshwater fishes

Environmental Parameters/ Season/ Growth Stage of Animal for Outbreak

- Grow at 15–35°C, pH 3.0–11.0, and in up to 1.0% NaCl

Treatment Measures

- 3.0 % NaCl for 1 h or 2.5 % NaCl for 2 h
- Exposure to 20 ppt seawater for 2 h and 30 ppt seawater for all dipping times (up to 3 h).

General Management Strategies for Fish Disease

The ways of prevention and contingent medical treatment of fish are very specific and often different from those of warm-blooded animals. They require a thorough knowledge of the environment of fish. Preventive arrangements consist of a complicated set of treatments elaborated on the base of a good knowledge of the etiology of disease and host (fish) biology. It concerns the elimination or restriction of infection (invasion) sources and the possibilities of its further expansion, as well as the enhancement of the condition of fish organisms in the way they are able to withstand the infection (invasion). Prevention is of basic importance in diseases elimination. No specific therapeutics have been developed for a number of diseases up to now, and the result of the application of effective, experimentally verified medicaments is often reversely affected by the operational conditions and/or the technology of rearing. Medical treatment has become economically unremunerative in this way.

In addition, some treatments cannot be performed in certain periods, e.g., in the growing season, during the wintering, or in some fish culture units (e.g., large ponds). That is why it is much more important to prevent the diseases than to recover them. The effective preventive treatments are to be applied above all in specialized fish culture units with closed warm water systems, in early fish fry rearing, hatcheries, trout farms, wintering ponds, and storage reservoirs.

Generally Accepted and Effective Principles are as Follows

Providing Water Sources Free of Pathogens

Underground waters are the most suitable water sources free of pathogens. These sources are limited both for trout farms and hatcheries and for other special fish culture units at present. The surface water from rivers and channels is used as the source of inflow water in most cases. In these situations, suitable filters can partially reduce the number of invasion stages of parasites in inflow water, above all when supplying smaller reservoirs with intensive culture. Bars are usually placed before these filters to separate rough particles. Sand filters consist of a set of sedimentation divisions terminated by a filter with fiber and sand. These types of filters catch above all the heavier parasite stages unable to move actively (e.g. spores). Lower efficiency is registered in the elimination of moving parasites like, e.g., infusorians. The water from the pond with fish stock is quite unsuitable for these purposes (esp. as the source of inflow water for trout farms, hatcheries, and units for early fish fry stages). Chemical treatment of inflow water is an emergency arrangement with often undesirable parallel effects. Disinfection of the water entering fish culture units by UV radiation is not still a usual way, although it can be considered a simple method of destroying viruses, bacteria, and moulds germs. Since the inflow water from rivers and channels is slightly turbid and contains a number of suspended solids and dissolved compounds, the disinfection efficiency of UV radiation is markedly reduced in these situations.

It is very profitable to supply the individual ponds and/or reservoirs independently, not throughflow. The water from each pond or reservoir should be drained separately and should not flow into any other. Especially quarantine ponds and other reservoirs can be separated in this way.

Protection From the Transfer of Pathogens

This principle means, above all, the transfer of pathogens by uncontrolled transport of fish and spawns. The transport of fish with unknown health conditions is to be avoided in principle. All transported fish are to be accompanied by veterinary certificate confirming that fish were examined before transporting them, they are healthy and originate from an environment in which no important transfer diseases appear. The list of these diseases is precisely stated in veterinary instructions. Except for the internal survey for each country also, the list of diseases stated in the international codex is obligatory for veterinary service. This list is currently specified with the development of diagnostic methods and improvement of knowledge about individual fish diseases. Some viral and bacterial diseases can be transferred also by spawns. Their transport must be completed by the same veterinary certificate as fish transport for this reason.

Fish introduced from other territories must be subjected to quarantine for one year regardless of whether they are native or extraneous species. The duration of quarantine can be prolonged, e.g., in the case of fish imported from abroad, for a period of 3 years. A prolonged quarantine period is of special importance, especially for spawners predestined for further reproduction of imported species. Self-sustaining stock production in individual farms and similar organizations is a significant way of preventing the dissemination of fish diseases. Only fish previously examined, free of diseases, and relevantly treated by medicinal baths are to be stocked into ponds and fish culture units. The stocking of fry originating from semi-artificial and artificial spawning not in contact with fish of higher age categories also minimizes the danger of infection.

The prevention of the introduction of coarse fish into ponds and fish culture units is another important arrangement that protects the stock against the transfer of pathogens. These fish are, above all, the source of ectoparasites, which are dangerous, especially in the period of decreased resistance of fish.

Aside from this, they can also transfer some other pathogens, which can result in heavy losses of important fish species. Adequate bars and filters can serve to prevent from coarse fish penetration. The protection of piscivorous birds from stepping into fish culture units (especially trout farms) is the prevention of limiting the expansion of some fish diseases. Protective nets are used to prevent the birds from running in. The numbers of piscivorous birds are regulated in localities where overpopulated.

Preventive control of snails (*Lymnaea* sp.) as intermediate hosts of some fish parasites can be performed by biological (introduction of black carp - *Myelopharyngodon piceus* or 3-years-old tench *Tinca tinca*), mechanical (placing nets in the inflow), physical (drying and freezing of the bottom) and chemical (application of molluscicides) ways. Safe and harmless removal of dead fish is a significant way how to prevent the further transfer of fish pathogens. Fresh or slightly decayed dead fish are decontaminated in the nearest veterinary facility. Lower masses of dead fish are to be burnt or buried into deep pits (approx. 2 m) at a distance of at least 20 m from the pond bank. The bottom of this pit and dead fish must be covered by burnt or chlorinated lime. The layer of at least 60 – 80 cm of the soil must cover the content of a pit.

Disinfection of Ponds, Fish Culture Units and Equipment; Winter Freezing and Summer Drying of Ponds

Disinfection is of great importance in the prevention and elimination of fish diseases. Preventive disinfection protects the fish stocks against pathogens. Hygiene of environmental conditions for fish is improved in this way. Focal disinfection is performed for control of the focus of dangerous fish disease. Natural physical phenomena are fully used for disinfection in intensive fish culture due to their economical convenience. It concerns the drying and freezing of the pond bottom. Most pathogens die after perfect drying of the pond bottom when its relative moisture has dropped by 10 – 15 %. The perfect freezing of the wet places and sun radiation (above all by its UV rays) have a very favorable effect on our conditions. The influence of these natural physical phenomena is exploited by summer drying and winter freezing of water reservoirs (ponds). Summer drying is a radical, long-term intervention during which all pathogens are controlled due to the perfect drying of the pond bottom. The aim of winter drying is to destroy the pathogens by freezing. It safely leads to the destruction of leeches (*Piscicola geometra*), fish lice (*Argulus* sp.), predatory larvae of water insects, eggs and spores of parasites, and also other pathogens. Employing natural methods for disinfection usually has a disadvantage in the long term (a number of months up to one year).

Chemical disinfection is an effective way of preventing and/or suppressing fish diseases. Usually, accessible disinfection preparations are used in fish culture (e.g., burnt lime, chlorinated lime, nitrogen lime, natrium hydroxide, potassium permanganate, formaldehyde, Chloramine, Chlorseptol, Jodonal, etc.). Burnt lime is mostly employed for disinfection of the bottom of ponds and reservoirs in the dose of 2.5 – 3 t. ha^{-1}, or chlorinated lime in the dose of 0.5 – 0.6 t. ha^{-1}. In the case of myxosporoses, nitrogen lime (5 t. ha^{-1}, or 0.5 $kg.m^{-2}$) is to be applied. Immediately after fishing out of the pond, disinfection of the fishing pit, pond ditches, and muddy wet places is performed on large ponds where whole-surface bottom disinfection is not possible. 5% water solution of formaldehyde, chlorinated lime (200 – 400 $mg.l^{-1}$), 0.5 % water solution of natrium hydroxide, Chloramine and chlorseptol (30 g. l^{-1}), or other disinfectants can be used for the treatment of concrete channels, troughs and other arrangements employed for fish culture. The same disinfectants and concentrations are to be used for the treatment of the equipment. Potassium permanganate (5 g. l^{-1}), Jodonal (2.8 – 4.5 ml. l^{-1}), and other disinfectants can also be employed for these purposes.

Optimization of Environmental Conditions

The optimization of natural environmental conditions is the main pre-condition for ensuring the good health condition of the stock during the rearing period. The following principles must be ensured:

- Optimal water quality without stressing physico-chemical effects. Keeping the oxygen concentration at an optimal level and protecting against water pollution is of special importance,
- Maximum development of natural food resources by adequate interventions, feeding fish by supplementary feed mixtures in sufficient amount and quality (attention should be paid to the quality of individual feed components and biofactors), basic preventive arrangements protecting the early developmental stages and young fish from bacteria and protozoans, including a sufficient amount of natural food of appropriate size and species composition,
- Responsible establishment of maximum stocking density. Inadequately high stocking density results in stress behavior worsened condition and resistance and makes the expansion of diseases easier. The stocking density is of special importance in trout farming and fish culture in special intensive units (but also in ponds),
- Prevention from stress situations evoked by other factors, manipulation during fishing out, transport, and long-term storage.

Regular control of health conditions and preventive treatment of fish

Preventive control of health conditions is to be carried out in hatcheries and early fry-rearing units twice a week and in highly productive intensified ponds, trout farms, and fish culture units, recycling warmed water weekly. Other stocks (esp. in usual pond culture) are investigated monthly. The health condition of fish is always to be controlled before fishing out, transporting fish, and stocking. Preventive treatment can be suggested based on the investigational results. This treatment is performed above all by the application of medicaments into the water environment and feeding by medicated feeds.

Other Preventive Principles

The ways of prevention from individual, most importantly viral, bacterial, fungal, and parasitic diseases, are described in adequate individual chapters. A specific, very effective way of preventing diseases is the vaccination of fish. Vaccines against the following relevant viral and bacterial diseases have recently been tested with different success: CCV, IPN, SVC, VHS, IHN, furunculosis, ERM, and vibriosis. Individual vaccines are applied intraperitoneally, personally, or in the form of a bath. Peroral application or bath are the most suitable ways from the point of view of fish culture practice. Also, vaccines against some other fish diseases, including parasitoses, are currently being developed.

Responsible Approach

- The transboundary movement of fry, fingerlings, and market-scale fish should minimize the risk of transmission of fish pathogens.
- Use of the Existing International Code of Practice dealing with transboundary movement and the use of introduced species in aquaculture as a framework for the formulation of regulations related to tropical latitudes and species. The existing database should be used to identify and determine risks from imported species.
- Reproductive history and disease history should maintain aquaculture stocks to promote disease-free breeding stocks and seed exchange in domestic and export.
- The aquaculture farm should use eggs or seeds free of specific pathogens (including eggs supplied to the backyard nursery)
- Governments, NGOs, and Regional agencies should promote outreach, extension, and technical support activities to raise farmers' and aquaculture industry's awareness of sanitation management issues, including Technical training, diagnostic services, and disease prevention and treatment.

- Fishermen, exporters, and farms should adopt effective fish health management measures to ensure fish health.

Aquaculture farms should take effective preventive measures to prevent, treat, and control diseases through the following methods:

- Reduce the pressure that may cause injury during handling and transportation,
- Implement good handling and storage practices for trash fish and artificial feed
- Establish effective and accurate diagnostic procedures
- Choose treatments that meet acceptable environmental and human health risks
- Evaluate management practices and identify key prevention areas
- Recognize the socio-economic impact: Adopting disease prevention, and treatment plans should offset the potential higher benefits.

General Problems in Fish Disease Diagnosis

There are several problems inherent in a disease diagnostic process. These have to be taken into account by advanced tools and methods to aid the diagnosis.

- No disease exhibits all the signs described in the literature. In most cases, the acute and chronic phases of a disease have differing signs. Therefore, the program has to be able to reach the right diagnosis with a partial set of signs.
- There is a time progression for every disease. A disease seen when the first clinical signs appear will exhibit different signs than when mortalities are already occurring.
- Since the program has to obtain input from a human user, the problem of terminology looms large. Despite efforts made by international organizations, no accepted vocabulary has been agreed upon for veterinary terminology. This is now changing with the incorporation of veterinary terminology in SNOMED. Moreover, cultural differences will also result in different terms being used for the same condition.
- In many cases, by the time the fish exhibit signs of a problem, there is already a secondary agent involved (virus+bacteria, fungus+bacteria, etc.). Therefore, the signs observed by the user may 'belong' to more than one disease in the program's database.
- 'It is human to err', but never more so than in our case. We have to take into account that the signs chosen by a user to describe a condition are

influenced by his knowledge and experience. Therefore, we have to deal with the possibility that 'wrong' signs will be entered by the user.

Conclusion

The success of fish production is directly proportional to the health of species. Hence, health monitoring and management cannot be ignored at any time. With little care, we can observe some of the abnormal signs, symptoms, or behavioral changes at the sites themselves. Whenever any such abnormality is noticed, expert advice should immediately be sought to avoid losses due to ill health and mortality. If the stock is kept healthy, returns will be rewarding. The rapid development of molecular biological techniques offers significant advantages for workers involved in fish disease diagnosis. Using nucleic acid as targets and new methods of analyzing polymorphism in this nucleic acid can improve specificity, sensitivity, and speed of diagnosis and offer means of examining the relationships between the genotype and phenotype of various pathogens. New hardware may make significant changes to the operation of these novel methods, either in the laboratory or in the field. Progress in techniques aids epidemiological studies as well as identifying causes of disease outbreaks or the presence of pathogens.

In recent years, the number of new publications describing new molecular techniques or methods has increased significantly. Such publications describe the development of new methods that appear very promising and useful. However, reports of the application of these techniques on a routine basis in diagnostic laboratories are few. In order for molecular biology to fulfill the promise of improved diagnosis and to be adopted by regulatory authorities, thorough trials of new methods are required and the results of these must be disseminated. Molecular methods have slowly established a place in the diagnosis of disease in aquaculture. The time has now come for their wide-scale adoption and application. Further funding is required to enable high-quality validation and trial of these methods as we move into the next century of fish health research and development.

Appendix

Table: Seasonal calendar for fish and shrimp diseases

Finfish diseases		
Diseases	Causative agent	Season/ growth stage of animal for outbreak
Bacterial diseases		
Gill rot	*Flexibacteria* spp *Pseudomonas* spp *Flavobacterium* spp *Aeromonas* spp	Spring and early summer months
Tail and fin Rot	*Aeromonas* spp *Pseudomonas* spp *Haemophilus* spp	Winter season
Edwardsiellosis	*Edwardsiella* tarda	Summer to post-monsoon
Dropsy	*Aeromonas hydrophila,* *Pseudomonas fluorescens*	Throughout the year
Aeromonasis	*Aeromonas hydrophila* *Aeromonar sobria* *Aeromonas caviae*	Spring to the early summer season
Columnaris	Flexibacter columnaris	Winter season
Fungal Diseases		
Branchiomycosis	*Branchiomyces Sanguinis*	Winter season
Saprolegniasis	*Saprolegnia parasitica* *Achyla* spp	
Epizootic ulcerative syndrome	*Aphynomyces invadans*	
Parasitic Diseases		
Argulosis	*Argulus* spp	Summer season
Larnaeasis	*Larnaea* spp	
Ergasilosis	*Ergasilus* spp	Late summer and autumn
Dactylogyrosis	*Dactylogyrus* spp	Late autumn or early winter
Gyrodactylosis	*Gyrodactylus* spp	

Viral Diseases		
TiLVD	Tilapia Lake virus	Late summer to monsoon
TiPVD	Tilapia Parvo virus	
	Shrimp diseases	
Diseases	Causative agent	Season/ growth stage of animal for outbreak
Bacterial diseases		
Vibriosis	Vibrio parahaemolyticus, V. harveyi, V. alginolyticus	Infection in late postlarval to early-to-late juvenile stages
Parasitic Diseases		
Hepatopancreatic microsporidosis	Enterocytozoon hepatopenaei	Infection in late postlarval to early-to-late juvenile stages
Viral Diseases		
WSSVD	White spot syndrome virus	All life stages
YHVD	Yellow head virus	Infection in late postlarval stages but mass mortality occurs in early-to-late juvenile stages

Fig. Seasonal calendar for fish diseases

Glossary (FAO)

Abscess: An aggregation of haemocytes (blood cells) associated with necrotic (decaying) host cells. Abscesses may or may not contain debris from invasive organisms which have been killed by host defences. In advanced abscesses there is a decrease in cell definition (especially the nuclei) towards the centre of the lesion, compared to cells around the periphery. Abscesses frequently involve breakdown of epithelial linings and may be surrounded by phagocytic and/or fibrocytic haemocytes.

Abiotic factors: Physical factors which affect the development/survival of an organism

Acquired immunity: Defence response developed following recovery from an infection (or vaccination) to a specific infectious agent (or group of agents)

Acute: Infection or clinical manifestation of disease which occurs over a short period of time (cf 'Chronic')

Adhesion: (Crustacea) binding of subcuticular tissues to the cuticle due to destruction of the cuticle by chitinolytic bacteria or fungi. This may impede moulting.

Aetiologic Agent (Etiologic): The primary organism responsible for changes in host animal, leading to disease

Aetiology (Etiology): The study of the cause of disease, including the factors which enhance transmission and infectivity of the aetiologic agent

Alevins: Fry of certain species of fish, particularly trout and salmonids that still have the yolk-sac attached

Anaemia: (Vertebrate) a deficiency in blood or of red blood cells

Anorexia: Loss of appetite

Antennal gland: (Crustacea) excretory pores at the base of the antennae (also known as kidney gland, excretory organ and green gland)

Antibody (Ab): A protein capable of cross-reacting with an antigen. In vertebrates, antibody is produced by lymphoid cells in response to antigens. The mechanism of antibody production in shellfish is not known.

Antigen: A substance or cell that elicits an immune reaction. An antigen may have several epitopes (surface molecules) to which antibody can bind (of Monoclonal and Polyclonal Antibodies)

Aquatic animals: Live fish, molluscs and crustaceans, including their reproductive products, fertilised eggs, embryos and juvenile stages, whether from aquaculture sites or from the wild

Aquaculture: Commonly termed "fish farming", it refers more broadly to the commercial hatching and rearing of marine and freshwater aquatic animals and plants

Ascites: Accumulation of serous fluid in the abdominal cavity; dropsy

Aseptic: Free from infection; sterile

Atrophy: Decrease in amount of tissue, or size of an organ, after normal growth has been achieved

Autolysis (-lytic): Enzyme induced rupture of cell membranes, either as a normal function of cell replacement or due to infection

Avirulent: An infection which causes negligible or no pathology (of Virulent)

Axenic culture: Culture containing cells of a single species (bacterial culture) or cell-type (tissue culture) (uncontaminated or purified)

Bacteriology: Science that deals with the study of bacteria

Bacteriophage: (Abbreviation - Phage) any virus that infects bacteria

Bacterium: (Bacteria) unicellular prokaryotic (nuclear material not contained within a nucleus) microorganisms that multiply by cell division (fission), typically have a cell wall; may be aerobic or anaerobic, motile or non-motile, free living, saprophytic or pathogenic

Basophilic: Acidic cell and tissue components staining readily with basic dyes (i.e. hematoxylin); chromatin and some secretory products in stained cells appear blue to purple

Bioassay: Aquantitative procedure that uses susceptible organisms to detect toxic substances or pathogens

Broodstock: Sexually mature fish, molluscs or crustaceans

Calcareous: Pertaining to or containing lime or calcium

Cannibalism: The eating of a species of animal by the same species of animal

Carrier: An individual who harbors the specific organisms of a disease without manifest symptoms and is capable of transmitting the infection; the condition of such an individual is referred to as carrier state

Ceroid: Non-staining metabolic by-product found in many bivalves. Abnormally high concentrations indicate possible environmental or pathogen-induced physiological stress

Chelating agent: Chemical agent used to decalcify calcium carbonate in mollusc shells or pearls, e.g., ethylenediaminetetracetic acid (EDTA)

Chemotherapeutant: Chemical used to treat an infection or non-infectious disorder

Chitin: Linear polysaccharide in the exoskeletons of arthropods, cell walls of most fungi and the cyst walls of ciliates

Chitinolytic (chitinoclastic): (Mycology and Bacteriology) chitin degrading organisms with enzymes capable of breaking down the chitin component of arthropod exoskeletons

Chronic: Long-term infection which may or may not manifest clinical signs

Clinical: Pertaining to or founded on actual observation

Chromatin: Nucleoprotein complex containing genomic DNA and RNA in the nucleus of most eukaryotic cells

Chromatophores: Motile, pigment-containing epidermal cells responsible for colour

Ciliostatic: Exotoxin toxin secreted by some bacteria that inhibits ciliary functions

Clone: A population derived from a single organism

Coagulation: Clotting (adhesion of haemocytes)

Conchiolin: Nitrogenous albuminoid substance, dark brown in colour, that forms the organic base of molluscan shells

Concretions: Non-staining inclusions in the tubule and kidney cells of scallops and pearl oysters, produced during the digestive cycle. Similar inclusions are also found in the gut epithelia of other bivalves.

Contagious: A disease normally transmitted only by direct contact between infected and uninfected organisms

Crustaceans: Aquatic animals belonging to the phylum Arthropoda, a large class of aquatic animals characterized by their chitinous exoskeleton and jointed appendages, e.g. crabs, lobsters, crayfish, shrimps, prawns, isopods, ostracods and amphipods

Cuticle: (Crustacea) the protein structure of arthropods consisting of an outer layer (epicuticle), an underlying exocuticle (pigmented), endocuticle (calcified) and membranous uncalcified layer. Chitin is in all layers except the epicuticle

Cyst: (a) a resilient dormant stage of a free-living or parasitic organism, or (b) a host-response walling off a tissue irritant or infection

Cytology: The study of cells, their origin, structure, function and pathology

Cytopathic effect: Pertaining to or characterized by pathological changes in cells

Decalcification: The process of removing calcareous matter

Decapitation: Cutting of the head portion

Deoxyribovirus: (DNA-virus) virus with a deoxyribonucleic acid genome (cf Ribovirus)

Diapedesis: Migration of haemocytes across any epithelium to remove metabolic byproduct, dead cells and microbial infections

Disease any deviation from or interruption of the normal structure or function of any part, organ, or system (or combination thereof) of the body that is manifested by a characteristic set of symptoms and signs and whose aetiology, pathology and prognosis may be known or unknown

Disease agent: An organism that causes or contributes to the development of a disease

Diagnosis: Determination of the nature of a disease

Disinfection: The application, after thorough cleansing, of procedures intended to destroy the infectious or parasitic agents of diseases of aquatic animals; this applies to aquaculture establishments (i.e. hatcheries, fish farms, objects that may have been directly or indirectly contaminated

DNA (ssDNA, dsDNA): Deoxyribonucelic acid. Nucleic acid comprised of deoxyribonucleotides containing the bases adenine, guanine, cytosine and thymine. Single strand DNA (ssDNA) occurs in some viruses (usually as a closed circle). In eukaryotes and many viruses, DNA is double-stranded (dsDNA)

DNA probes: Segments of DNA labelled to indicate detection of homologous segments of DNA in samples of tissues or cultures (see RNA probes)

Dropsy: The abnormal accumulation of serous fluid in the cellular tissues or in a body cavity

Ecdysal gland: (Crustacea) see Y-organ

Ectoparasite: A parasite that lives on the outside of the body of the host

ELISA: Enzyme Linked Immunosorbent Assay, used to detect antigen (antigen capture ELISA) or antibody (antibody capture ELISA)

Emaciation: a wasted condition of the body

Endemic: Present or usually prevalent in a population or geographical area at all times

Endothelia: Pertaining to or made up of endothelium

Endothelium: The layer of epithelial cells that lines the cavities of the heart and of the blood and lymph vessels, and the serous cavities of the body originating from the mesoderm

Endosymbiosis: An association between two organisms (one living within the other) where both derive benefit or suffer no obvious adverse effect

Envelope: (Virology) lipoprotein membrane composed of host lipids and viral proteins (non-enveloped viruses are composed solely of the capsid and nucleoprotein core)

Enzootic: Present in a population at all times but, occurring only in small numbers of cases

Eosinophilic: Basic cell and tissue components staining readily with acidic dyes (i.e. eosin); stained cells appear pink to red

Epibiont: Organisms (bacteria, fungi, algae, etc.) which live on the surfaces (of fouling) of other living organisms

Epipodite: (Crustacea) cuticular extension of the base (protopodite) of the walking legs (pereiopods)

Epitope: The component of an antigen which stimulates an immune response and which binds with antibody

Epizootic: Affecting many animals within a given are at the same time; widely diffused and rapidly spreading (syn. Epidemic - used for human disease)

Epidemiology: Science concerned with the study of the factors determining and influencing the frequency and distribution of disease or other health related events and their causes in a defined population for the purpose of establishing programs to prevent and control their development and spread

Epizootiology: The study of factors influencing infection by a pathogenic agent

Epithelium: The layer of cells covering the surface of the body and all gastrointestinal linings. Epithelia are usually one cell thick and supported by a basal membrane

Epitope structural: Component of an antigen which stimulates an immune response and which binds with antibody

Erosion: Destruction of the surface of a tissue, material or structure

Eukaryotean: Organism that contains the chromosones within a membrane-bound nucleus (of Prokaryote)

Exoenzyme: Extracellular enzyme released by a cell or microorganism

Exopthalmia: Abnormal protrusion of the eyeballs

Exoskeleton: (Crustacea) the chitin and calcified outer covering of crustaceans (and other arthropods) which protects the soft-inner tissues

Exudate: Material, such as fluid, cells, or cellular debris, which has escaped from blood vessels and has been deposited in tissues or on tissue surfaces, usually as a result of inflammation

Euthanasia: An easy or painless death

Filtration: Passage of a liquid through a filter, accomplished by gravity, pressure or vacuum (suction)

Finfish: Fresh or saltwater fish of any age

Fry: Newly hatched fish larvae

Fingerling: A young or small fish

Fixation: Preservation of tissues in a liquid that prevents protein and lipid breakdown and necrosis; the specimen is hardened to withstand further processing; and the cellular and sub-cellular contents are preserved in a manner close to that of the living state

Fixative: A fluid (e.g. aldehyde or ethanol-based solutions)) that prevents denaturation and autolysis by cross-linking of proteins

Foreign bodies: Any organism or abiotic particle not formed from host tissue

Formalin: A 37% solution of formaldehyde gas

Fouling: The mass colonisation of hard substrates by free-living organisms. Extreme fouling of living organisms, such as molluscs or shrimp, can impede their normal body-functions leading to weakening and death

Fungus: Any member of the Kingdom Fungi, comprising single-celled or multinucleate organisms that live by decomposing and absorbing the organic material in which they grow oyster farms, shrimp farms, nurseries), vehicles, and different equipment.

Gaping: Weakened molluscs that cannot close their shells when removed from water; this rapidly lead to desiccation or predation of the soft-tissues and is indicative of molluscs in poor condition (including possible infection)

Gram's Stain: Stain used to differentiate bacteria with permeable cells walls (Gram-negative) and less permeable cell walls (Gram-positive)

Granulomas: Any small nodular delimited aggregation of granular haemocytes, or modified macrophages resembling epithelial cells (epithelioid cells)

Granulomatosis: Any condition characterized by the formation of multiple granulomas

Granulosis virus: Baculoviridae belonging to subgroup (B), characterised by a single nucleocapsid within an envelope. Granulosis viruses form intra-nuclear ellipsoid or rounded occlusion bodies (granules or capsules) containing one or two virions

Gross signs: Signs of disease visible to the naked eye

Haematopoietic: Pertaining to or effecting the formation of blood cells

Haematopoietic tissue: (Decapoda) a sheet of tissue composed of small lobules surrounded by fibrous connective tissue which lies along the dorso-lateral surfaces of the posterior portion of the cardiac stomach (Brachyura) or surrounding the lateral arterial vessels, secondary maxillipeds and epigastric tissues (Penaeidae and Nephropidae); (Bivalves) unknown; (Vertebrates) spleen

Haemocytes: Blood-cells

Haemolymph: Cell-free fraction of the blood containing a solution of protein and non-proteinaceous defensive molecules

Haemocyte infiltration: Accumulation of haemocytes around damaged or infected tissues; since the type of haemocytes most commonly responsible for phagocytosis are granulocytes, focal infiltration is often referred to as a "granuloma"

Haemocytopenia: A reduction in the number of cells in the circulatory system, usually associated with a reduction in blood-clotting capability

Haemocytosis: Systemic destruction of blood cells (syn. Haemolysis)

Haemorrhage: (Vertebrate) escape of blood from the vessels; bleeding (Invertebrate) uncontrolled loss of haemocytes due to tissue trauma, epithelial rupture, chronic diapedesis

Hatcheries: Aquaculture establishments raising aquatic animals from fertilized eggs

Hepatopancreas: Digestive organ composed of ciliated ducts and blind-ending tubules, which secrete digestive enzymes for uptake across the digestive tubule epithelium; also responsible for release of metabolic by-products and other molecular or microbial wastes (of Metaplasia, Diapedesis)

Histology: The study that deals with the minute structure, composition and function of tissues

Histolysis: Breakdown of tissue by disintegration of the plasma membranes

Histopathology: Structural and functional changes in tissues and organs of the body which cause or are caused by a disease seen in samples processed by histology

Host: Individual organism infected by another organism

Homogenate: Tissue ground into a liquid state in which all cell structure is disinter grated

Husbandry: Management of captive animals to enhance reproduction, growth and health

Hyperplasia: Abnormal increase in size of a tissue or organ due to an increase in number of cells

Hypertrophy: Abnormal enlargement of cells due to irritation or infection by an intracellular organism

Hyphae: (Mycology) tubular cells of filamentous fungi; may be divided by cross-walls (septae) into multicellular hyphae, may be branched. Inter-connecting hyphae are called mycelia

Icosahedra: Shape of viruses with a 5-3-2 symmetry and 20, approximately equilateral, triangular faces

IFAT: Indirect Fluorescent Antibody Test/Technique; a technique using unlabelled antibody and a labelled anti-immunoglobulin to form a 'sandwich' with any antigen-bound antibody

Immunity: Protection against infectious disease conferred either by the immune response generated by immunization or previous infection or by other non-immunologic factors

Immunization: Protection against disease by deliberate exposure to pathogen antigens to induce defence system recognition and enhance subsequent responses to exposure to the same antigens (syn Vaccination)

Immunoassay: Any technique using the antigen-antibody reaction to detect and quantify the antigens, antibodies or related substances (see ELISA, IFAT, DFAT)

Immunodepression: Decrease in immune system response to antigens due to an infection (same or different agent) or exposure to an immunosuppressant chemical.(syn. Immunosupression)

Immunofluorescence: Any immuno-histochemical method using antibody labeled with a fluorescent dye. Direct - if a specific antibody or antiserum with a fluorochrome and used as a specific fluorescent stain. Indirect - if the fluorochrome is attached to an antiglobulin, and a tissue constituent is stained using an unlabeled specific antibody and the labeled antiglobulin, which binds the unlabeled antibody

Immunoglobulin (Ig): Family of proteins constructed of light and heavy molecular weight chains linked by disulphide bonds; usually produced in response to antigenic stimulation

Immunohistochemistry: Application of antigen-antibody interactions to histochemical techniques, as in the use of immunofluorescence

Immunology: Branch of biomedical science concerned with the response of the organisms to antigenic challenge, the recognition of self and not self, and all the biological (in vivo), serological (in vitro), and physical chemical aspects of immune phenomena

Immunostimulation: Enhancement of defense responses, e.g., with vaccination

Immunization: Induction of immunity

Inclusion body: Non-specific discrete bodies found within the cytoplasm or nucleus of a cell. Frequently viral (cf Cowdry body, Polyhedrin Inclusion / Occlusion Bodies), or bacterial microcolonies (cf RLOs) (syn. Inclusions)

Infectious: Capable of being transmitted or of causing infection

Infection: Invasion and multiplication of an infectious organism within host tissues. May be clinically benign (cf sub-clinical or 'carrier') or result in cell or tissue damage. The infection may remain localized, subclinical, and temporary if the host defensive mechanisms are effective or it may spread an acute, sub-acute or chronic clinical infection (disease)

Infiltration: (Invertebrates) haemocyte migration to a site of tissue damage or infection by a foreign body/organism ('inflammation'). Infiltration may also occur for routine absorption and transport of nutrients and disposal of waste products

Inflammation: (Vertebrate) initial response to tissue injury characterised by the release of amines which cause vasodilation, infiltration of blood cells, proteins and redness that may be associated with heat generation

(Invertebrates) infiltration response to tissue damage or a foreign body. The infiltration may be focal, diffuse or systemic (syn. Infiltration)

Innate immunity: Host defence mechanism that does not require prior exposure to the pathogen

Intensity of infection: The number of infectious agents in an individual organism or specimen; "mean" intensity is the average number of infectious agents present in all infected individuals in a sample

Intercellular: Situated or occurring between the cells in a tissue

Interstitial tissue: Tissue or cells between epithelial bound organ systems; also known as (cells) Leydig tissue (molluscs) or connective tissue

take different forms and be associated with saprobionts (bacterial, fungal or protistan) proliferation

Notifiable Diseases: 'Diseases notifiable to the OIE' means the list of transmissible diseases that are considered to be of socio-economic and/ or public health importance within countries and that are significant in the international trade in aquatic animals and aquatic animal products.

Nuclear Polyhedrosis Virus (NPV): Baculoviruses (Type A) which produce intranuclear polyhedral protein matrices

Nucleocapsid: Protein-nucleic acid complex which may form the core, capsid and/or helical nucleoprotein of the virion

Occlusion: (Vascular) filling or blocking of vascular sinuses by haemocytes; (perivascular) infiltration of haemocytes, several cells deep into the tissues surrounding vascular sinuses; (luminal) filling or blocking of gonoducts, renal ducts, digestive tubules or ducts by haemocytes or other cell debris

Oedema (edema): Presence of abnormally large amounts of fluid in the intercellular spaces of the body

Opportunistic: Organism capable of causing disease only when a host's resistance is pathogen lowered by other factors (another disease, adverse growing conditions, drugs, etc.)

Osmoregulation: Maintenance of osmolarity by a simple organism or body cell with respect to the surrounding medium

Outbreak: The sudden onset of disease in epizootic proportions

Overt: Open to view; not concealed

Parasite: An organism which lives upon or within another living organism (host) at whose expense it obtains some advantage, generally nourishment

Parasitology: Science that deals with the study of parasites

Passage (Virology) the successive transfer of a virus or other infectious agent through a series of experimental animals, tissue culture, or synthetic media with growth occurring in each medium

Patent infection: Period when clinical signs and/or the infectious organism can be detected (cf Prepatent)

Pathogen: An infectious agent capable of causing disease

Pathogenicity: The ability to produce pathologic changes or disease

Pathognomonic: Sign or symptom that is distinctive for a specific disease or pathologic condition

Pathology: Deals with the essential nature of disease, especially of the structural and functional changes in tissues and organs of the body which cause or are caused by a disease

PCR: Polymerase Chain Reaction, a process by which nucleic acid sequences can be replicated ('nucleic acid amplification')

Pereiopods: (Crustacea) thoracic appendages ('walking legs') (cf Pleopods and Uropods)

Periostracum: (Molluscs) calcareous layers of shell which may contain quinine-tanned protein

Phagocytosis: Uptake by a cell of material from the environment by invagination of its plasma membrane

Plasma membrane: Trilaminar membrane enclosing the cytoplasm and organelles of a cell

Pleiopod: Small legs of some crustaceans

Pleomorphic: Organism demonstrating more than one body form within a life-cycle

Polyadenalated RNA: Messenger RNA (mRNA) which has a polyadenylate sequence bound to the 3' end of the molecule. This is common in most eukaryote mRNA and is present in some riboviruses. The function of this addition is unknown

Polyclonal antibodies (PAb): (More correctly, but rarely, termed 'Polyclonal antiserum') an antiserum prepared from an organism exposed to an antigen. The PAb contains several different antibodies, each specific to a different epitope of the same antigen. (see Monoclonal antibody)

Polyhedral Inclusion/ Occlusion Body (POB, PIB): Protein-based crystalline matrix made up of Polyhedrin (Baculovirus group A - Nuclear Polyhedrosis Viruses (NPV)) or Granulin (Baculovirus group B - Granulosis Viruses (GV)). Baculovirus group C do not form occlusion bodies

Polymorphic: (a) capability of molecules, such as enzymes, to exist in several forms; (b) ability of nuclei of certain cells (e.g., haemocytes) to change shape; and (c) ability of microorganisms to change shape (e.g., in different host species or tissues)

Pop-eye: Abnormal protrusion of the eyes from the eye sockets

Postlarvae (PL): The stage following metamorphosis from larvae to juvenile in the life cycle of Crustacea. In penaeid shrimp, this is commonly counted in days after appearance of postlarval features, e.g., PL12 indicates a post-larvae that has lived 12 days since its metamorphosis from the zoea stage of development

Predator: An organism that derives elements essential for its existence from organisms of other species, which it consumes and destroys

Predispose: to make susceptible to a disease which may be activated by certain conditions, as by stress

Preening: (Crustacea) cleaning surface tissues or eggs exposed to fouling (cf Epibionts and Fouling); some crustaceans have modified appendages to enhance preening (e.g., the gill-rakers of Brachyura)

Prepatent period: Period between infection and the manifestation of clinical or detectable signs of disease

Prevalence: Percentage of individuals in a sample infected by a specific disease, parasite or other organism

Prokaryote: (syn. Bacteria) cellular micro-organisms in which the chromsones are not enclosed within a nucleus

Prophylactic (-axis): Action or chemotherpeutant administered to healthy animals in order to prevent infection

Pustule: A sub-epidermal swelling containing necrotic cell debris as a result of inflammation (haemocyte infiltration) in response to a focal infection

Putative: Signifies that which is commonly thought, reputed or believed

Pyknosis/Pyknotic: Contraction of nuclear contents to a deep staining (basophilic) irregular mass, sign of death cell (cf Karryorhexis and Karyolysis)

Quarantine: Holding or rearing of aquatic animals under conditions which prevent their escape, and the escape of any pathogens they may be carrying, into the surrounding environment. This usually involves sterelisation/disinfection of all effluent and quarantine materials. Quarantine measures are measures developed as a result of risk analysis to prevent the transfer of disease agents with live aquatic animal move ments, with pre-border, border and post-border health management processes, however, such activities are equally applicable to intra-national movements of live aquatic animal

Repair: Process to re-establish anatomical and functional integrity of tissues after an injury or infection

Reservoir: (Host or infection) an alternate or passive host or carrier that harbors pathogenic organisms, without injury to itself, and serves as a source from which other individuals can be infected

Resistance: (To Disease) (cf Acquired immunity and Innate immunity) the capacity of an organism to control the pathogenic effects of an infection. Resistance does not necessarily negate infection ('Refraction') and varying degrees of tolerance to the infection may be manifest. Heavy sub-clinical infections are indicative of resistance (syn. Tolerance; opp. Susceptible)

Resistance: (Antibiotic or 'drug' resistance) the capability of a microbe to evade destruction by an antibiotic. This may arise from changes in the antigenic properties of the microbe. Survival and multiplication lead to development of drug resistant strains of the pathogen. This may confer resistance to related (heteroresistance) or non-related antibiotics (multiple drug resistance)

Ribosomes: Intracytoplasmic granules which are rich in RNA and function in protein synthesis

Risk: The probability of negative impact(s) on aquatic animal health, environmen tal biodiversity and habitat and/or socio-economic investment(s)

RNA: Ribonucleic acid consisting of ribonucleotides made up of the bases (ssRNA, dsRNA) adenine, guanine, cytosine and uracil

RNA probes: Segments of RNA which are labelled to detect homologous segments of RNA or DNA in tissue or culture samples (cf DNA probes)

rRNA: (Ribosomal RNA) RNA component of the ribonucleoprotein organelle responsible for protein synthesis within a cell

Saprobionts: (Syn. Saprotroph) organisms which obtain nutrition from dead organic matter

Schizonts: The multinucleated stage or form of development during schizogony

Secondary: Infection infection resulting from a reduction in the host's resistance as a consequence of an earlier infection

Septicaemia: Systemic disease associated with the presence and persistence of pathogenic microorganisms or their toxins in the blood; blood poisoning

Serology: Term now used to refer to the use of such reactions to measure serum antibody titers in infectious disease (serologic tests), to the clinical correlations of the antibody titer (the 'serology' of a disease) and the use of serologic reactions to detect antigens

Serum: Fluid component of coagulated haemolymph

Shipment: A group of aquatic animals or products thereof destined for transportation

Sporangium: (Mycology) hyphal swelling which contains motile or non-motile zoospores; release is via a pore or breakdown of the sporangial wall. (syn. Zoospo rangium)

Sporangium: (Bacteriology) the cell, or part of a cell, which subsequently develops into an endospore (intracellularly formed spore)

Spore: Infective stage of an organism that is usually protected from the environ ment by one or more protective membranes (syn. Zoospores)

Sporogenesis: Formation of or reproduction by spores; sporulation

Sterilization: Any process (physical or chemical) which kills or destroys all contaminating organisms, irrespective of type; a sterile environment (aquatic or solid) is free of any living organism

Stress: The sum of biological reactions to any adverse stimuli (physical, internal or external) that disturb the organism's optimum operating status

Sub-clinica: (Asymptomatic) an infection with no evident symptoms or clinical signs of disease, or a period of infection preceding the onset of clinical signs (cf Prepatent)

Surveillance: A systematic series of investigations of a given population of aquatic animals to detect the occurrence of disease for control purposes, and which may involve testing of samples of a population

Susceptible: An organism which has no immunity or resistance to infection by another organism

Syndrome: An assembly of clinical signs which when manifest together are indicative of a distinct disease or abnormality (syn. Pathognomic/ Pathognomonic)

Synergistic: (Infection) pathology increased by two or more infections by different agents, compared with the effect from individual effects (opp. to 'antago nistic' or 'suppressive', where one infection counteracts the other)

Systemic: Pertaining to or affecting the body as a whole

Systemic infection: An infection involving the whole body

Tail rot: Disintegration of tail and fin tissue

Telson: (Crustacea) terminal segment of the abdomen which overlies the uropods

Tomont: The non-feeding, dividing stage or form in the life cycle of certain protozoa that typically encysts and produces tomites by fission

Transmission: Transfer of an infectious agent from one organism to another

Horizontal - direct from environment (e.g., via ingestion, skin and gills)

Vertical - prenatal transmission (i.e., passed from parent to egg); may be either inside the egg (intra-ovum) or through external exposure to patho gens from the parent generation

Transport: Movement of stocks between locations by human influence

Trauma: An effect of physical shock or injury

Treatmentaction: Taken to eradicate an infection (cf Prophylaxis)

Trophozoites: The active, motile, feeding stage of a protozoan organism, as contrasted with the non-motile encysted stage

Tumour: Abnormal growth as a result of uncontrolled cell division of a localised group of cells

Ubiquitous: Existing or being everywhere

Ulcer: Excavation of the surface of an organ or tissue, involving sloughing of necrotic inflammatory tissue

Uropods: (Crustacea) the terminal appendages underlying the telson that form the 'tail fan' (see Pereiopods and Pleopods)

Vaccine: An antigen preparation from whole or extracted parts of an infectious organism, which is used to enhance the specific immune response of a susceptible host

Vacuolated: Containing spaces or cavities within the cytoplasm of a cell

Veliger: (Mollusc) ciliated planktonic larval stage

Velum (Velar): (Mollusc) ciliated feeding surface of veliger larvae

Viable: Capable of living or causing a disease

Virion: Individual viral particlecontaining nucleic acid (the nucleoid), DNA or RNA (but not both) and a protein shell, or capsid

Virogenic stroma(e): Site of viral replication or assembly (syn. Viroplasm)

Virogensis: Production of virions

Virology: Branch of microbiology which is concerned with the study of viruses and viral diseases

Virulence: The degree of pathogenicity caused by an infectious organism, as indicated by the severity of the disease produced and its ability to invade the tissues of the host; the competence of any infectious agent to produce pathologic effects; virulence is measured experimentally by the median lethal dose (LD_{50}) or median infective dose (ID_{50})

Virus one of a group of minute infectious agents, characterized by a lack of independent metabolism and by the ability to replicate only within living host cells

Y-organ: (Crustacea) (syn. Ecdysal gland) gland resonsible for production of the moulting hormone ecdysone. Production of the moulting hormone is controlled by a moult inhibiting hormone synthesised in the eye-stalk

Zoea larvae: (Crustacea) stage following metamorphosis from the nauplius larva, characterised by four pairs of thoracic appendages; may be referred to as protozoea where differentiation between the nauplius and mysis or postlarva stage of development is difficult

Zoospores: Motile, flagellated and asexual spores